Lecture Notes in Biomathematics

Managing Editor: S. Levin

37

Tomoko Ohta

Evolution and Variation of Multigene Families

Springer-Verlag
Berlin Heidelberg New York 1980

Author

Tomoko Ohta
National Institute of Genetics
1,111 Yata,
Mishima-shi, Shizuoka-ken
411 Japan

AMS Subject Classifications (1980): 92-02, 92A10

ISBN 3-540-09998-0 Springer-Verlag Berlin Heidelberg New York
ISBN 0-387-09998-0 Springer-Verlag New York Heidelberg Berlin

Printing and binding: Beltz Offsetdruck, Hemsbach/Bergstr.
2141/3140-543210

Lecture Notes in Biomathematics

Managing Editor: S. Levin

37

Tomoko Ohta

Evolution and Variation
of Multigene Families

Springer-Verlag
Berlin Heidelberg New York

Lecture Notes in Biomathematics

FOREWORD

During the last decade and a half, studies of evolution and variation have been revolutionized by the introduction of the methods and concepts of molecular genetics. We can now construct reliable phylogenetic trees, even when fossil records are missing, by comparative studies of protein or mRNA sequences. If, in addition, paleontological information is available, we can estimate the rate at which genes are substituted in the species in the course of evolution. Through the application of electrophoretic methods, it has become possible to study intraspecific variation in molecular terms. We now know that an immense genetic variability exists in a sexually reproducing species, and our human species is no exception.

The mathematical theory of population genetics (particularly its stochastic aspects) in conjunction with these new developments led us to formulate the "neutral theory" of molecular evolution, pointing out that chance, in the form of random gene frequency drift, is playing a much more important role than previously supposed. I believe that the traditional paradigm of neo-Darwinism needs drastic revision. Also, the importance of gene duplication in evolution, as first glimpsed by early Drosophila workers, has now been demonstrated by directly probing into genetic material.

Recently, it has been discovered that some genes exist in large-scale repetitive structures, and that they are accompanied by newly described phenomena such as "coincidental evolution". Working out the population genetical consequences of multigene families is a fascinating subject, for which Dr. Ohta has been largely responsible. Prior to her theoretical work in this field, she had already made many original contributions to theoretical population genetics, particularly to problems relating to evolution and variation at the molecular level.

In this monograph, Dr. Ohta presents, in an organized form, various work that she has done on evolution and variation of multigene families. Some of this work has been published in scattered journals. I am sure that this monograph will be useful not only to mathematical biologists who are interested in evolution, but also to students of evolution who are ambitious enough to try to understand the nature of progressive evolution in molecular terms.

Motoo Kimura

PREFACE

This monograph is written so that several fragmentally reported analyses on multigene families may be presented in the form of an organized theory. It is also intended to bring various seemingly unrelated subjects together into a unified understanding. These subjects include antibody diversity, molecular evolution and probability of gene identity.

It is my great pleasure to express my appreciation to Dr. Motoo Kimura, from whom I learned most of the necessary procedures to carry out the analyses, and who has given me stimulating discussion and encouragement throughout the course of this work. I also express my gratitude to Dr. Susumu Ohno and Dr. Bruce S. Weir for going over the English from the beginning to the very end, and providing many suggestions which greatly improved the presentation. Thanks are also due to Dr. Jack L. King, Dr. Yoshio Tateno and Dr. Naoyuki Takahata for their many valuable suggestions and comments on the manuscript, and to Mrs. Yuriko Isii and Mrs. Sumiko Yano for their excellent typing.

Tomoko Ohta

INTRODUCTION

In evolution of higher organisms, gene duplication has apparently
played a very important role. For more complex organization, more
genetic information is needed, and gene duplication seems to be the
only way to achieve it. Ohno (1970) emphasized this theme from various theoretical and empirical points of view. Among better known
protein molecules, there are many examples which have originated by
gene duplication; e. g., myoglobin and hemoglobin α as well as β,
lysozyme and α-lactalbumin, trypsin and chymotrypsin, immunoglobulin
κ, λ and heavy chains etc. All these examples clearly show that gene
duplication and subsequent functional differentiation have been a
major feature of evolution.

Tandemly repeated genes which form a multigene family such as
ribosomal RNA genes, transfer RNA genes or immunoglobulin genes represent extreme cases of gene duplication. They are not simply the
result of a single duplication and subsequent differentiation but have
arisen through many cycles of duplications and periodical reductions
of gene members during the long course of evolution. The genetic
organization of these multigene families has been greatly clarified
in recent years by the remarkable progress of molecular biology (e. g.
Hood et al. 1975, Tartof 1975). The evolutionary establishment of
such families involves two related processes: the origin of new variation, and its spread within populations. Thus, it is important to
interpret the new knowledge of multigene families based upon a sound
population genetics theory, for it may provide a clue to understand
the progressive evolution of higher organisms, thereby giving us a new
insight on how genetic information has accumulated in the course of
evolution. In this monograph, I shall present my population genetics
approach on the evolution and variation of multigene families. The
analyses cover the several different disciplines of biology such as
population genetics, molecular biology and immunology. Thus, the
monograph is intended to be of interest to scientists in broad areas
of modern biology.

CHAPTER 1

ORGANIZATION AND PROPERTIES OF ANTIBODY GENES
AND OTHER MULTIGENE FAMILIES

1.1 Multigene family

A multigene family is defined as a group of genes or nucleotide
sequences with the following characteristics (Hood et al. 1975):
multiplicity, close linkage, sequence homology, and related or over-
lapping phenotypic functions. Figure 1.1 shows a diagram of a multigene family. In the figure, G_1 or G_2 represents a repeating unit
(gene or nucleotide sequence) and is tandemly arranged on the same
chromosome. Hood et al. (1975) classified multigene families into
three categories; simple-sequence, multiplicational, and informational
multigene families. Table 1.1 shows examples and properties of the
three categories. The first category, the simple sequence multigene
family is often called satellite DNA or highly repeated DNA which may
be isolated from the bulk of DNA by buoyant density centrifugation.
As its name implies, this category is very highly repeated and the
repeating units are simple and homogeneous. The function, if any, of
simple sequence multigene families is not known. It is generally
considered that they are genetically inert, although they may have
some role for chromosomal organization. To the second class, the
multiplicational multigene family, belong several very important gene
families such as histone, ribosomal RNA and transfer RNA genes. They
are often repeated several hundred times in a genome; the more or less
fixed gene member characterizing each species. The gene products of
this category have very fundamental biological roles.

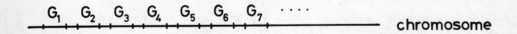

Figure 1.1   Diagram showing the structure of a multigene
family (from Hood et al. 1975, slightly
modified).

Table 1.1 Classification and properties of multigene families

Category	Gene products	Multiplicity	Gene or protein homology	Information content	Examples of coincidental evolution	Examples of change in family size
Simple-sequence satellites	None known	10^3-10^6	80-100%		Different satellites of Drosophila	Mouse satellite
Multiplicational 18S-28S RNA	RNA, protein	100-600	97-100%	One unit	Spacer regions of X. laevis and X. mulleri	Gene number in X. mulleri and X. laevis
5S RNA		100-1200	97-100%	One unit	Spacer regions of X. laevis and X. mulleri	Gene number in X. mulleri and X. laevis
tRNA		6-400		One unit		
histones		10-1200	87-99%	Few units	Histone mRNAs of two species of sea urchin	Gene number in different sea urchin species
Informational antibodies	Proteins	100S	30-100%	Many units	Rabbit and mouse κ chains	V_λ and V_κ in mammals
hemoglobins		~10	<75-100%	Few units	Human and cow δ chains	Human and rabbit β-like genes

In all instances characteristics refer to a single closely linked set of genes. From Hood et al. (1975).

High multiplicity probably reflects the fact that organisms need these
gene products in a large quantity at certain developmental stages.
Although individual genes are nearly identical within each family, the
spacers between the genes are heterogeneous in their lengths (Wellauer
et al. 1974). Furthermore, some internal repetitions are often found
in the spacer region and the spacer seems to have undergone faster
evolutionary changes than the gene itself (Brown et al. 1971).

The last category, the informational multigene family, is most
interesting from the evolutionary view point. Yet, the immunoglobulin
gene families constitute the only well known examples of this class.
It is possible that, in the near future, some structural protein genes,
such as keratin or collagen genes, may be shown to form multigene
families, although they may not approach the complexity of immunoglobulin multigene families. In fact, much progress has been made on
elucidating the organization of chorion genes of silkworms which can
be interpreted to form a small informational multigene family (Kafatos
et al. 1977, Goldsmith and Basehoar 1978). The informational multigene families are stimulating subjects of research, because they have
the potential to generate a great variety of related but individually
diverse information. In somatic recombinations between variable
region genes drawn from a large pool and constant region genes drawn
from a much smaller pool, one finds the remarkably efficient mechanism
of generating great multitudes of specific antibodies within individual organisms (e. g., Marx 1978). Needless to say, the above
requires the expression of individual genes within this class of
multigene family to be clonal in somatic cells, as we shall shortly
see. This clonal expression is in sharp contrast to the ubiquitous
expression that characterizes members of the multiplicational multigene family. In the following, I shall review the organizational
characteristics of antibody genes.

1.2 Antibody gene families

Origin of antibody diversity is one of the most stimulating subjects in modern biology (e.g. Proc. Cold Spring Harbor Symp. Vol 41).
There are three major classes of immunoglobulins: I_gG, I_gM and I_gA.
The most common class of adult mammals is I_gG and the individual molecule
of this class is composed of two identical light chains and two identical heavy chains. Figure 1.2 shows a diagram of the basic structure
of I_gG. A light chain has two homologous domains, each having about 110
amino acids, and one intrachain disulfide bridge. Each domain is
recognized as a unit by its pattern of folding. The various domains

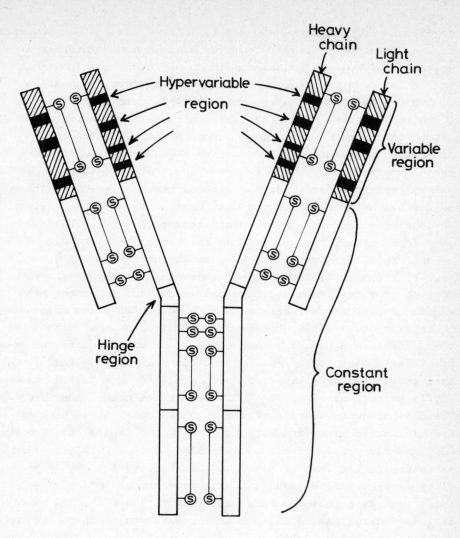

Figure 1.2 A model of the structure of the immunoglobulin (I_gG) molecule. According to
Capra and Edmundson (1977), there are
four hypervariable regions in the heavy
chain, however, some investigators
apparently assign only three regions as
hypervariable.

are recognizably homologous to one another because of distant sequence
similarities; apparently they are the result of ancient internal gene
duplications. The amino terminal domain is different from chain to
chain and is called variable region. The carboxy terminal half is
stable, although several types are known, and is called the constant
region. As to the heavy chain, the total size is about twice as long
as the light chain and it has four homologous domains. The amino
terminal domain is the variable region and the constant region contains three domains. There is also a small hinge region at the middle
of the heavy chain constant region. Various classes of immunoglobulins are defined by the kind of constant region of the heavy chain.

The light chains occur in two distinct forms, κ (kappa) and λ
(lambda). The κ chain, the λ chain and the heavy chain are each located on different chromosomes and thus constitute different gene
families. Figure 1.3 illustrates the situation. Thanks to the recent
remarkable progress of immunogenetics and molecular immunology, the
fine details of gene arrangements within these families are being un-
raveled (Seidman et al. 1978, Tonegawa et al. 1978). The variable
region gene family has recently been found to be divided into two gene
families; V (variable) region and J (junction) region gene families,
the latter containing only about 15 amino acids (Figure 1.4) (Tonegawa
et al. 1978, Weigert et al. 1978). Although exact gene number in a
genome is not known for V and J region gene families, it is guessed
that several hundred genes are present in a V gene family (Seidman et
al. 1978), and five or so in a J gene family. The mouse λ gene family
is an exception so far examined and, in this family, the sequence
variability is very limited and hence the gene number in a family is
much less. According to Tonegawa et al. (1977), there are only two
genes in this family. As to the constant region genes, there are only
a few genes in one genome. During development, V region, J region and
the constant region are somatically joined together to make a complete
light or heavy chain. In as much as the expression of light or heavy
chain genes made complete by somatic recombination is clonal as already noted, each malignant myeloma cell line which is of monoclonal origin synthesizes only species each of light and heavy chains.
The so called allelic exclusion (haploid expression) also contributes
to the monoclonal expression. It is a remarkable system to increase
antibody diversity through combinatorial use of genetic information.
V, J and the constant region are said to be united by "combinatorial
joining" while association of a light chain and a heavy chain is
called "combinatorial association" (Weigert et al. 1978). Both are

Germ cell nucleus

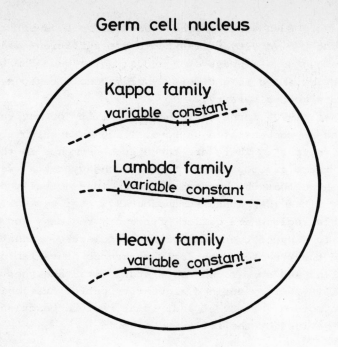

Figure 1.3   A model showing the chromosomal location
of the three immunoglobulin gene families
(from Watson 1976, slightly modified).

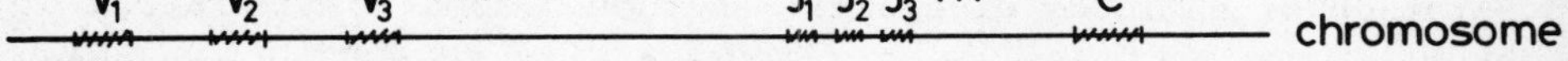

Figure 1.4   A model illustrating the organization of
a gene family of the light chain.

somatic processes. Such somatic processes would not be efficient in
increasing antibody diversity unless there are sufficient genetic
variabilities among gene members of V and J region gene families.
Next I shall review amino acid diversity of variable regions which
directly reflects gene diversity.

     Wu and Kabat (1970) found that the variability of the variable
region sequences is concentrated on several segments which are called
"hypervariable regions". There are three such segments in the light
chain, whereas there are four segments in the heavy chain (see Fig.
1.2). It was shown that the hypervariable regions actually constitute
antigen binding sites (Capra and Edmundson 1977, for review). The
non-hypervariable regions are called "framework regions" and are
needed for the folding of the immunoglobulin molecule. Kabat et al.
(1976) compiled the reported amino acid sequences of variable regions
of immunoglobulins and summarized their statistical analyses on
sequent variability. They defined a quantity called "variability" to
represent amino acid diversity at individual residue positions. The
variability is defined by the following formula,

$$\text{Variability} = \frac{\text{number of different amino acids occurring at a given position}}{\text{frequency of the most common amino acid at that position}} \ .$$

Hypervariable regions were found by using this quantity (Wu and Kabat
1970). The Figure 1.5 shows some examples how variability changes at
the framework and at the hypervariable regions.

     A very stimulating controversy yet unsettled is whether or not
somatic mutations of the variable region contribute significantly to
the generation of antibody diversity. Although somatic generation of
antibody diversity by combinatorial joining and association has been
verified by the molecular studies, the problem of somatic mutations in
terms of amino acid substitution has not yet been resolved (Weigert et
al. 1978). A simple somatic mutation hypothesis assumes a certain special mutational mechanism which preferentially affects the hypervariable regions during ontogeny (Jerne 1977, Leder et al. 1977). A more
recent somatic mutation hypothesis seems to take some intermediate
position in which hypervariability is attributed both to evolutionary
accumulations of mutations and somatic mutations (Weigert and Riblet
1977, Weigert et al. 1978). On the other hand, certain investigators
apparently dismiss somatic mutations as of negligible significance to
the generation of antibody diversity (e. g. Seidman et al. 1978).

ALL HUMAN KAPPA LIGHT CHAINS

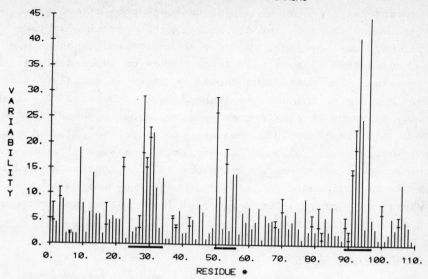

ALL HUMAN LAMBDA LIGHT CHAINS

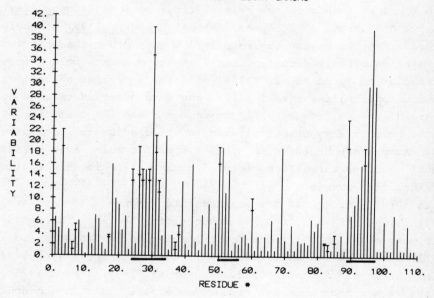

Figure 1.5   Two examples of "variability" of immuno-
             globulins (from Kabat et al. 1976).
             Hypervariable regions are underlined.

An important approach that has been neglected in such studies is to analyse sequence variation as a part of the gene pool of evolving species. For example, even inbred mouse lines preferentially used to generate monoclonal myelomas by these investigations such as NZB or BALB/c are represented by a very large number of individuals at the present, thus, such an inbred line should be regarded as an evolving population, in spite of the intense inbreeding in its past. A variant haplotype may arise by fairly frequent unequal crossing-overs in the region of the gene family. Thus, one may expect polymorphism within the inbred line. Accordingly, the quantitative analyses of sequence variabilities become necessary by using population genetics methods. In chapter six of this monograph, I shall present my analyses on sequence variability of variable regions of immunoglobulins from the standpoint of population genetics.

1.3 Coincidental evolution

In general, multigene families have two remarkable characteristics in evolution. One is that the number of repeating genes in one family changes during evolution. An extreme example is the numbers of 5S ribosomal RNA genes in two related frog species; Xenopus laevis has about 24,000 5SrRNA gene copies whereas Xenopus mulleri has only 9,000 genes or so (Brown and Sugimoto 1973). The other characteristic of the evolution of multigene families is the phenomenon called coincidental evolution (Hood et al. 1975), which is sometimes called horizontal evolution. As the term implies, the same mutation spreads to all members of the multigene family to become a species characteristic. Figure 1.6 presents a hypothetical example of coincidental evolution. An actual example may be found in species specific amino acid residues found within various variable regions of human κ-type immunoglobulin light chains, for example.

It is awfully difficult and unnatural to imagine that the same mutation independently affected all members of a family. Thus, some mechanism must exist that spreads the one mutation horizontally to nearly all members of the multigene family during the course of speciation (Brown and Sugimoto 1973, Hood et al. 1975, Tartof 1975). Several hypotheses have been put forward to explain this phenomenon. The model of saltatory replication suggests that certain genes are suddenly replicated in the genome of a species, thereby creating the species-specific multigene family (Southern 1970, Walker 1971, Davidson and Britten 1971). According to this hypothesis, the coincidental evolution takes place in a relatively short period of time.

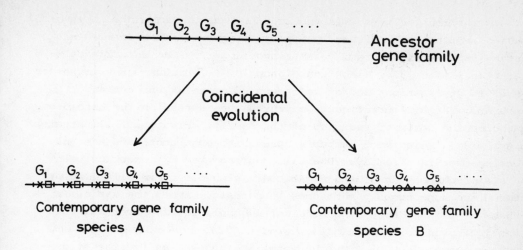

Figure 1.6   A diagram illustrating the coincidental
              evolution from a common ancestor gene family
              (from Hood et al. 1975, slightly modified).

A second hypothesis, the model of gene conversion, postulates that the
sequence correction of multiple gene copies occurs within each multigene family. An extreme form of this model is the master-slave hypothesis (Callan 1967) in which the slave gene copies of the last
generation are eliminated, and a new set of slaves in the exact image
of the master is created de novo in each generation. A library hypothesis (Salser et al. 1976) is a modification of this model in which
certain library genes are kept in the genome for the reference and,
once in a while, a new set of gene copies are made de novo for the
reference gene. A still more modified hypothesis suggests that the
gene correction occurs using any gene member of a multigene family as
the reference (Gally and Edelman 1972). Under the master-slave model,
the coincidental evolution takes place in every generation, and therefore, all the gene copies of the same family are identical at the
start of each individual's life. By contrast, gene members of a given
multigene family may become progressively heterogeneous under the
model in which coincidental evolution occurs slowly over the period
of many generations. Such slow horizontal spread of the same mutation
within a given multigene family is most compatible with the third
hypothesis.

This third hypothesis is the model of unequal crossing-over which assumes that unequal but homologous crossing-overs occasionally affecting each multigene family are responsible for the coincidental evolution (Smith 1974, Black and Gibson 1974). Unequal crossing-overs result in duplications and/or deletions of certain gene copies. Therefore, certain mutant genes may by chance spread on the chromosome to become the dominant sequence of the family. Figure 1.7 illustrates an example in which gene number 4 spreads through continued unequal crossing-over in a family. Unequal crossing-over may occur between the sister-chromatids at the somatic division of a germ cell line or between the homologous chromosomes at meiosis. Among the various hypotheses explained here, the model of unequal crossing-over has been investigated in most detail (e. g. Smith 1974, 1976, Ohta 1976, 1979, Perelson and Bell 1977). Furthermore, it thus far enjoys the strongest experimental support (e. g. Wellauer et al. 1976, Musich et al. 1978, Fedoroff and Brown 1978, Yamamoto and Miklos 1978). In the next chapter, the model of unequal crossing-over will be reviewed and discussed in detail.

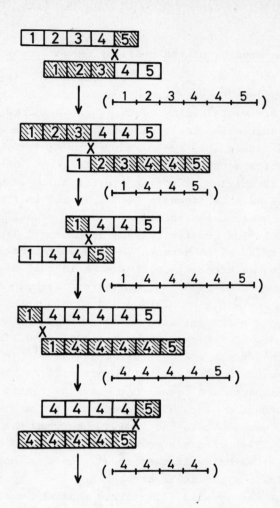

Figure 1.7 A model of gene fixation by
unequal crossing-over.

CHAPTER 2

MODELS OF UNEQUAL CROSSING-OVER

2.1 Simulation studies of the model of unequal crossing-over

Smith (1974) as well as Black and Gibson (1974) were the first to
show that the diversity of antibody genes and other multigene families
may be adequately explained by assuming unequal crossing-overs. They
used computer simulation for understanding the process of coincidental
evolution. The basic idea is to start from linearly arranged genes,
every one of which is represented by a different integer from others
so that it is identified as a gene lineage in a later period of simulation (as in the example given in Fig. 1.7) and to find out how gene
members become homogeneous through repeated unequal crossing-overs.
Thus their simulation treats a single chromosomal line and an unequal
crossing-over takes place between the sister-chromatids, i. e. intra-
chromosomal unequal crossing-over at somatic cell division. In order
to simulate inter-chromosomal unequal crossing-over at meiosis, one
needs to treat the whole population and thus an experiment of a much
greater scale becomes necessary.

In the model of Black and Gibson (1974), the shift of genes of
unequal matching at a crossing-over is always one gene and duplications and deletions occur by equal chance of 1/2. They examined the
loss of gene lineages and found that when the family size is large,
more unequal crossing-overs are needed to reach a certain level of
homogeneity of genes within the family than when the family size is
small. Smith (1974) used a somewhat more realistic model in which the
shift of unequal matching at the crossing-over is not restricted to
one gene but follows a certain probability law. He set an allowed
latitude of the family size, i. e., the shift at unequal crossing-
overs is determined by a random number so that the family size fits
within a given allowed latitude. For example, the allowed latitude of
10% of the family size means that the family size randomly varies in
the range $n_0 - n_0/10 \sim n_0 + n_0/10$, where n_0 is the initial number of
genes in the family. He counted the number of unequal crossing-overs
needed until a majority of the genes in the family becomes a single
lineage; i. e., the expansion of one gene lineage. He called this
number "crossover fixation time" since the process is analogous to

fixation of a selectively neutral mutant in finite populations due to random genetic drift. The analogy will be extended in the later section of this chapter.

Smith also found that the crossover fixation time becomes longer as the family size gets larger. As to the allowed latitude, the larger the allowed latitude, the smaller is the crossover fixation time. However, beyond a certain value, increasing the allowed latitude becomes ineffective in accelerating the crossover fixation. For example, when the average family size is 100, the crossover fixation is more rapid for the case of the allowed latitude, 95 ∿ 105, than the case of 99 ∿ 101. However, the former is not different from the case of 75 ∿ 125. Smith considers that the allowed latitude of 10% of the family size is adequate and, in such cases, the crossover fixation time is roughly 20 times that of the family size. Smith further proposes that, by this model, the evolution of all categories of multigene families (see Tab. 1.1) may be satisfactorily explained. When the crossover fixation is rapid in relation to the appearance of new mutations, the genes are expected to become homogeneous in the family, whereas when the crossover fixation is slow, the gene members should become heterogeneous in the family. Here, one factor which should not be forgotten in such discussions must be noted. The point is that the genetic variation involves the Mendelian population as a whole, therefore, analysis of a single chromosomal line is not enough for understanding the extent of gene diversities in the family. The population genetics approach on this problem will be presented in the next chapter.

2.2 Diffusion approximation of coincidental evolution

Before proceeding to the population genetic approach, I shall present diffusion approximation of the process of coincidental evolution by intra-chromosomal (sister-chromatid) crossing-overs (Ohta 1976). The model is the same as that of Smith (1974) or Black and Gibson (1974), in which a single chromosomal line is undergoing continuous unequal crossing-overs. In order to make an analytical treatment possible, I assume that the shift at an unequal crossing-over is by one gene and that duplications and deletions occur alternately in cycles. Thus, in this model, the gene number per family stays the same, n_g. Figure 2.1 illustrates the model. Let us call two consecutive crossing-overs involving duplication and deletion "one cycle" of the process.

Let $x_{i,k}$ be the frequency of the k-th gene lineage in the i-th

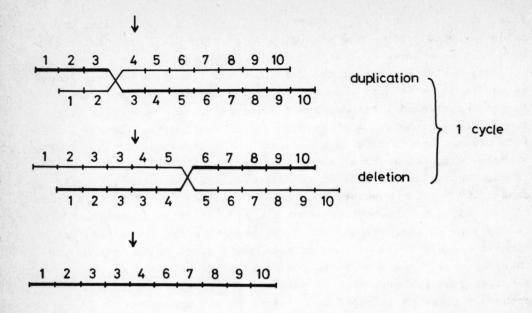

Figure 2.1   Diagram illustrating the model of
intra-chromosomal (sister-chromatid)
unequal crossing-over.

family. If one starts from the linearly arranged n_g genes each representing a different lineage as explained in the previous section,
$x_{i,k}$ = $1/n_g$ at the start. I shall treat the change of $x_{i,k}$ by unequal crossing-overs by using the diffusion model which has been
developed by Kimura (1964) for analysing the gene frequency change in
a finite population due to random genetic drift. To apply the diffusion model, one needs to evaluate the mean ($M(\Delta x_{i,k})$), the variance
($V(\Delta x_{i,k})$) and the covariance ($W(\Delta x_{i,k}\Delta x_{i,\ell})$) of the change of $x_{i,k}$
per unit time. Therefore let us evaluate these quantities per one
cycle of unequal crossing-overs. Table 2.1 shows the derivation. At
the expansion phase, $x_{i,k}$ increases by the amount $1/n_g$ with probability $x_{i,k}$ since a random member of the family is assumed to duplicate.
Similarly, $x_{i,\ell}$ increases with probability $x_{i,\ell}$. The first three
columns show these relationships. At the deletion phase, $x_{i,k}$ and
$x_{i,\ell}$ decrease by the amount $1/n_g$ with probabilities $x_{i,k}$ and $x_{i,\ell}$
respectively. Actually, the expansion or deletion by one gene is not

Table 2.1 Derivation of the mean and variance of the change of gene lineage frequencies by intra-chromosomal (sister-chromatid) unequal crossing-over.

Duplication phase			Deletion phase			Total		
Change of $x_{i,k}$	Change of $x_{i,\ell}$	Probability	Change of $x_{i,k}$	Change of $x_{i,\ell}$	Probability	Change of $x_{i,k}$	Change of $x_{i,\ell}$	Probability
$+\frac{1}{n_g}$	0	$x_{i,k}$	$-\frac{1}{n_g}$	0	$x_{i,k}$	0	0	$x_{i,k}^2$
			0	$-\frac{1}{n_g}$	$x_{i,\ell}$	$+\frac{1}{n_g}$	$-\frac{1}{n_g}$	$x_{i,k}x_{i,\ell}$
			0	0	$1-x_{i,k}-x_{i,\ell}$	$+\frac{1}{n_g}$	0	$x_{i,k}(1-x_{i,k}-x_{i,\ell})$
0	$+\frac{1}{n_g}$	$x_{i,\ell}$	$-\frac{1}{n_g}$	0	$x_{i,k}$	$-\frac{1}{n_g}$	$+\frac{1}{n_g}$	$x_{i,\ell}x_{i,k}$
			0	$-\frac{1}{n_g}$	$x_{i,\ell}$	0	0	$x_{i,\ell}^2$
			0	0	$1-x_{i,k}-x_{i,\ell}$	0	$+\frac{1}{n_g}$	$x_{i,\ell}(1-x_{i,k}-x_{i,\ell})$
0	0	$1-x_{i,k}-x_{i,\ell}$	$-\frac{1}{n_g}$	0	$x_{i,k}$	$-\frac{1}{n_g}$	0	$x_{i,k}(1-x_{i,k}-x_{i,\ell})$
			0	$-\frac{1}{n_g}$	$x_{i,\ell}$	0	$-\frac{1}{n_g}$	$x_{i,\ell}(1-x_{i,k}-x_{i,\ell})$
			0	0	$1-x_{i,k}-x_{i,\ell}$	0	0	$(1-x_{i,k}-x_{i,\ell})^2$

$$M(\Delta x_{i,k}) = 0, \quad V(\Delta x_{i,k}) = \frac{2}{n_g^2} x_{i,k}(1-x_{i,k}) \quad \text{and} \quad W(\Delta x_{i,k}, \Delta x_{i,\ell}) = \frac{-2}{n_g^2} x_{i,k}x_{i,\ell}$$

exactly equal to the frequency change by $1/n_g$, however since expansion and deletion occur by cycle, the change may be equated to $1/n_g$. The last three columns of the table give the change of $x_{i,k}$ and $x_{i,\ell}$ by one cycle of unequal crossing-overs with the corresponding probabilities. From them, $M(\Delta x_{i,k})$, $V(\Delta x_{i,k})$ and $W(\Delta x_{i,k} x_{i,\ell})$ may be calculated as follows.

$$M(\Delta x_{i,k}) = 0$$

(2.1)

$$V(\Delta x_{i,k}) = 2x_{i,k}(1 - x_{i,k})/n_g^2$$

and

$$W(\Delta x_{i,k} \Delta x_{i,\ell}) = -2x_{i,k}x_{i,\ell}/n_g^2 .$$

A very interesting analogy may be noted between the above formulae and the corresponding equations for the change of gene frequency in a finite population by random drift. If y_1 and y_2 are the frequencies of selectively neutral mutant alleles at a locus in a population with effective size N_e, the mean, the variance and the covariance of changes of y_1 and y_2 in one generation are given in the standard population genetics text book (e. g. Crow and Kimura 1970 p. 330); $M_{\Delta y_1} = 0$, $V_{\Delta y_1} = y_1(1 - y_1)/2N_e$ and $W_{\Delta y_1 \Delta y_2} = -y_1 y_2/2N_e$. Thus, one cycle of unequal crossing-overs is equivalent to $4N_e/n_g^2$ generations of random genetic drift. Once the analogy becomes clear, then the established theories in population genetics may be applicable. Therefore, the crossover fixation time discussed in the previous section may be directly obtained by replacing N_e with $n_g^2/4$ in Kimura and Ohta's (1969a) formula which gives mean fixation time of a mutant in the population by random drift. It becomes

$$t_1(p) = -\frac{1}{p} \{n_g^2(1 - p)\log_e(1 - p)\} \approx n_g^2$$

(2.2)

where p is the initial frequency of a gene lineage and equal to $1/n_g$ which is very small. The above formula gives the mean of the crossover fixation time. It is further possible to extend the analogy to the formula for the variance of the crossover fixation time (Kimura and Ohta 1969b). We get

$$\text{var}(t_1) \approx 0.286n_g^4 .$$

These analytical solutions are based on several simplifying assumptions. Later I shall consider more general cases and also extend
this diffusion approximation method to the level of breeding populations.

2.3 Coincidental evolution as a birth-death process

As before, we consider a single chromosomal line and start from
the linearly arranged genes, each representing a different gene lineage. There are N_0 genes and, therefore N_0 gene lineages, at the
beginning. The unequal crossing-over (sister-chromatids) is assumed
to occur by shift of one gene but duplication and deletion do not
necessarily occur in alteration. Thus the gene number may change with
time. We denote by N_t the number of genes at time t (in number of
unequal crossing-overs). Perelson and Bell (1977) investigated such
a process by applying the birth-death process. Let $P_{n_i}(t)$ be the
probability that there are n_i genes of the i-th gene lineage at time
t. If we assume that the point of unequal crossing-over is randomly
determined and that the duplication and deletion occur with equal
probability of 1/2, the change of $P_{n_i}(t)$ by one unequal crossing-over
is given by ,

$$P_{n_i}(t + 1) = (1 - \frac{n_i}{N_t})P_{n_i}(t)$$

$$+ \frac{1}{2}\left\{\frac{n_i - 1}{N_t} P_{n_i-1}(t) + \frac{n_i + 1}{N_t} P_{n_i+1}(t)\right\}$$

$$\text{for } n_i \geq 1 \qquad (2.4)$$

and

$$P_0(t + 1) = P_0(t) + \frac{1}{2}\frac{1}{N_t} P_1(t) .$$

Perelson and Bell (1977) approximated the process by replacing
N_t with its mean value, N_0, and obtained the solutions of the above
equations by using the result of Feller (1957). They are written as,

$$P_0(t) = \frac{t}{t + 2N_0}$$

and $\qquad (2.5)$

$$P_{n_i}(t) = \frac{(t/2N_0)^{n_i-1}}{(1 + t/2N_0)^{n_i+1}} \qquad n_i \geq 1 .$$

Also, the mean of crossover fixation time can be obtained. Let $M_e(t)$ be the mean number of extinct genes at time t. It may be expressed

$$M_e(t) \;=\; N_0 P_0(t) \;=\; \frac{N_0 t}{t + 2N_0} \tag{2.6}$$

When $M_e(t) = N_0 - 1$, only one gene lineage is present in the family, i. e. one gene lineage is fixed. Therefore, by solving the equation

$$N_0 - 1 \;=\; \frac{N_0 t}{t + 2N_0}$$

for t, we get the mean crossover fixation time by the birth-death process,

$$T_{bd} \;=\; 2N_0(N_0 - 1) \tag{2.7}$$

(Perelson and Bell 1977). Since this is the mean number of unequal crossing-overs until fixation of a gene lineage, it is about twice as large as the number of cycles of unequal crossing-over until fixation obtained by the diffusion approximation in the previous section.

CHAPTER 3

POPULATION GENETICS APPROACH

3.1 Model

The analyses in the previous chapter concerned themselves only
with the evolution of a single chromosomal line through sister-chromatids unequal crossing-over at mitotic division of a germ cell line.
Actually, however, genes are exchanged between individuals by sexual
reproduction in higher organisms. Thus the population genetics approach
is needed for understanding the evolution and variation of multigene
families. The process becomes very complicated, therefore, in
the following, I use a number of simplifications. Also, the simplest
and most convenient quantity called "identity coefficient" will be
investigated which is defined as the probability of two genes belonging to the family being identical by descent. The identity coefficient
is equivalent to the measure of homozygosity in population genetics
which is so often used for studying genetic variation within a population and between related species at enzyme loci (Nei 1975).

We assume a randomly mating population with an effective size N_e.
The gene families in the population may be different from each other,
but they are mutually related through common ancestors. It is assumed
that the gene family is evolving by the means of mutations, inter-
chromosomal crossing-overs (equal or unequal) at meiosis, intra-
chromosomal unequal crossing-overs during mitotic divisions as well as
by random genetic drift. Thus one generation consists of the following four processes (Table 3.1). Let v be the mutation rate per generation. In this chapter, I shall use v as the rate per gene but in
the later chapters, v is used for a smaller unit inside a gene such
as per amino acid site or codon. It is assumed that whenever a mutation occurs, the result represents a new character rather than
resembling one of the pre-existing alleles (infinite allele model of
Kimura and Crow 1964). Let β be the rate per family of equal inter-
chromosomal crossing-over (with no shift), γ' be that of unequal
inter-chromosomal crossing-over with the mean shift of m' genes, and
γ, that of unequal intra-chromosomal crossing-over with the mean shift
of m genes. In order to simplify, I assume that the mean family size,
n_g, remains unchanged and all parameters, v, $1/N_e$, β, γ and γ' are
much less than unity.

Table 3.1

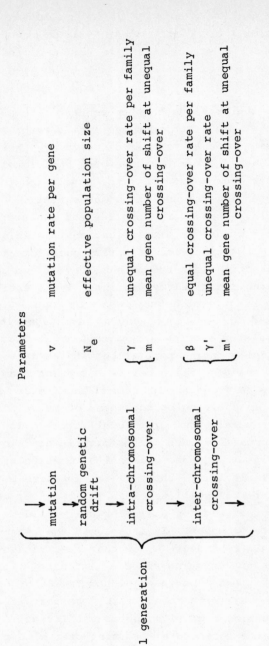

	Parameters	
mutation	v	mutation rate per gene
random genetic drift	N_e	effective population size
intra-chromosomal crossing-over	γ	unequal crossing-over rate per family
	m	mean gene number of shift at unequal crossing-over
inter-chromosomal crossing-over	β	equal crossing-over rate per family
	γ'	unequal crossing-over rate
	m'	mean gene number of shift at unequal crossing-over

1 generation

The evolving population of multigene families may be diagrammed by the following Figure 3.1. Family members may be grouped into a number of family types. Here a family type is a group of genetically identical gene families. Each family type may also be diagrammed as shown in the figure; n_g genes are linearly arranged and genes may be grouped into several distinct gene lineages. Let p_i be the frequency of the i-th gene family type in the <u>population</u> and $x_{i,k}$ be the frequency of the k-th gene lineage in the i-th <u>family type</u>.

The simplest way of analysing this complicated system is to treat the changes of both p_i and $x_{i,k}$ by the diffusion approximation. This is possible under certain simplifying assumptions. In the next section, I shall present my analyses on this double diffusion approximation. Later more exact treatments will be given.

Random mating population

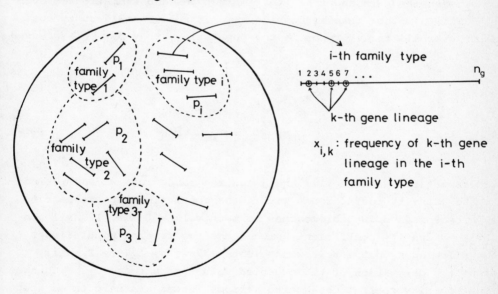

Figure 3.1 Diagram showing a population of a
multigene family.

3.2 Approximate analyses by double diffusion process

The identity coefficient is defined as the probability of two randomly chosen genes belonging to the same multigene family being identical by descent. Let us define an average identity coefficient at two different levels: the identity coefficient of genes on one chromosome <u>versus</u> that between members of the same multigene family residing on different chromosomes in a population. In terms of gene lineage frequencies, $x_{i,k}$, the two measures of the identity coefficient may be expressed as follows (Ohta 1978a), just like the measures of homozygosity and gene identity at ordinary gene loci,

$$c_i = \sum_k x_{i,k}^2 \tag{3.1}$$

and

$$c_{ij} = \sum_k x_{i,k} x_{j,k} \tag{3.2}$$

where c_i is the identity coefficient of the i-th family type, c_{ij} is that between the i-th and the j-th family types and summation is over all k. If I further take in consideration the average identity coefficient over all family types in the population, they may be expressed as follows.

$$C_{w1} = \sum_i c_i p_i \tag{3.3}$$

and

$$C_{w2} = \sum_i \sum_j c_{ij} p_i p_j \tag{3.4}$$

where summation is over all i and j and the subscripts w1 and w2 indicate the identity coefficients within the population with respect to one chromosome and two chromosomes respectively. Later, I shall consider the identity coefficient between the two related populations.

Following my previous reports (Ohta 1978a, b, 1979), I shall present the derivation of the expected values of C_{w1} and C_{w2}. Changes of the identity coefficient due to various forces are calculated step by step under simplifying assumptions. The following vector is defined.

$$C = \begin{bmatrix} C_{w1} \\ C_{w2} \end{bmatrix} . \tag{3.5}$$

The mutation decreases the value of C by the rate, 2v, in each generation, we therefore have,

$$C' = (1 - 2v)C . \qquad (3.6)$$

By sampling, C_{w1} does not change, whereas C_{w2} changes by inbreeding due to a finite population with effective size N_e. Thus, after sampling we have,

$$C'' = \begin{bmatrix} 1 & 0 \\ \dfrac{1}{2N_e} & 1 - \dfrac{1}{2N_e} \end{bmatrix} C' . \qquad (3.7)$$

From the formula (2.1), the change of C_{w1} through one cycle of unequal intra-chromosomal crossing-overs, is given by the following equation,

$$E(\Delta C_{w1}) = E\{\sum_k (x_{i,k} + \Delta x_{i,k})^2 - \sum x_{i,k}^2\}$$

$$= E\{\sum_k (\Delta x_{i,k}^2)\} = \frac{2}{n_g^2}(1 - C_{w1})$$

in which E denotes taking expectation over the family type frequencies in the population. C_{w2} does not change by unequal intra-chromosomal crossing-over. One unequal crossing-over with shift of m genes is equivalent to m/2 cycles of unequal crossing-over, provided that there exists no defined linear order among multiple gene copies on the chromosome. This condition will be examined in detail in chapter 4. If an unequal intra-chromosomal crossing-over occurs by the rate γ with mean shift of m genes in each generation, the change of the vector C" in one generation becomes,

$$C''' = \begin{bmatrix} 1 - \alpha & 0 \\ 0 & 1 \end{bmatrix} C'' + \begin{bmatrix} \alpha \\ 0 \end{bmatrix} \qquad (3.8)$$

in which $\alpha = m\gamma/n_g^2$.

Inter-chromosomal crossing-overs consist of equal (with rate β) and unequal (by rate γ' with the mean shift of m' genes) ones. In order to make an analytical treatment possible, the simplifying assumption of randomness in the linear arrangement of multigenes on the chromosome is again needed.   Therefore, the identity coefficient is assumed to be independent of the chromosomal distance between individual genes.   More exact analyses by Kimura and Ohta (1979) will be given later to clarify the adequacy and limitation of the present approach due to this assumption.   The advantage of the approximate analyses is that not only the identity coefficient but also some other quantities such as higher moments may be evaluated which will be given later.

I shall start from calculation of the effect of <u>equal</u> crossing-over under the simplifying assumption.   Now consider that one cross-ing-over takes place between one of the i-th and one of the j-th family types as given by the following figure.   The crossing-over point divides the gene family into two parts r and 1 - r.   In the

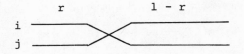

Figure 3.2   Diagram illustrating the inter-
chromosomal crossing-over.

newly formed recombinant family, the frequency of the k-th gene line-age is the weighted average of the i-th and j-th family types.   By letting a recombinant family be the i'-th type, the new gene lineage frequency becomes

$$x_{i',k} = r\, x_{i,k} + (1 - r) x_{j,k} \ . \tag{3.9}$$

The identity coefficient of the recombinant family may be expressed as follows.

$$c_{i'} = \sum_{k} x_{i',k}^{2} = r^{2} c_{i} + (1 - r)^{2} c_{j} + 2r(1 - r) c_{ij} \tag{3.10}$$

Similarly, by letting another recombinant family be the j'-th type,

$$c_{j'} = r^{2} c_{j} + (1 - r)^{2} c_{i} + 2r(1 - r) c_{ij} \tag{3.11}$$

and

$$c_{i'j'} = r(1 - r)(c_i + c_j) + \{r^2 + (1 - r)^2\}c_{ij} . \qquad (3.12)$$

Let us assume here that r is uniformly distributed between 0 and 1. Therefore, taking into consideration an expected change of the identity coefficient, we may put $E(r) = 1/2$, $E(r^2) = 1/3$ and so forth. By using the formulas (3.9 - 3.12), the expected identity coefficient change due to inter-chromosomal equal crossing-over in one generation may be obtained which becomes,

$$C^{(IV)} = \begin{bmatrix} 1 - \dfrac{\beta}{3} & \dfrac{\beta}{3} \\[2ex] \dfrac{\beta}{6N_e} & 1 - \dfrac{\beta}{6N_e} \end{bmatrix} C''' . \qquad (3.13)$$

When both β and $1/N_e$ are much less than the unity, the above equation becomes approximately,

$$C^{(IV)} \approx \begin{bmatrix} 1 - \dfrac{\beta}{3} & \dfrac{\beta}{3} \\[2ex] 0 & 1 \end{bmatrix} C''' . \qquad (3.14)$$

The coefficient of C_{w2}, $\beta/6N_e$, is derived from the consideration that the left part of the recombinant chromosome (part r of Fig. 3.2) has $2N_e - 1$ ways of matching with the right part (part $1 - r$ of Fig. 3.2) of any other chromosome in the population, therefore the first term of the right hand side of the formula (3.12) becomes by taking the expectation in one generation,

$$\beta \times (\frac{1}{2N_e - 1}) \times E\{r(1 - r)\} \times \{E(c_i) + E(c_j)\} \approx \frac{\beta}{6N_e} C_{w1} .$$

On the other hand C_{w2} decreases by the amount, $(\beta/6N_e)C_{w2}$, since the second term of the right hand side of the equation (3.12) is $1 - 2r(1 - r)$ and the term $2r(1 - r)$ results in the same coefficient as the above, $(\beta/6N_e)$.

Next, let us proceed to the inter-chromosomal unequal crossing-over. To begin with, one needs to calculate the mean and the variance of gene lineage frequency changes in a unit time in the manner as was

the case with intra-chromosomal unequal crossing-overs (see Table 2.1).
Since the covariance of gene lineage frequency changes is also needed
in applying the diffusion equation method in the later section, I
shall here complete the calculations for deriving the mean, the variance and the covariance. At the moment, in order to simplify the
calculation, let us assume that the duplication and deletion by one
gene occur alternately in cycle consequent to individual unequal
crossing-overs (Ohta 1979). Let us consider the situation where one
cycle of an unequal crossing-over has taken place between one of the
i-th and one of the j-th family types, as in the following Fig. 3.3.
In this case, n_g = 10, and the points of exchange are between 3 and
4 in the i-th type and between 2' and 3' in the j-th type. Thus, both
3 and 3' are included in the new family with 11 gene units. This is
the expansion phase and is followed by the deletion of one gene (contraction phase). Expansion and deletion phases constitute one cycle
as in the case of sister-chromatid unequal crossing-over (Fig. 2.1).

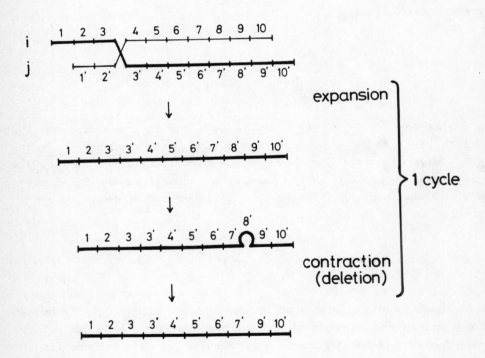

Figure 3.3 Diagram illustrating the model of inter-
chromosomal unequal crossing-over, for
deriving the mean, the variance and the
covariance of the change of gene lineage
frequencies.

Changes of C_{w1} and C_{w2} are calculated for this expansion and contraction process. In addition to this process, chromosomal segments are exchanged. In the above example (Fig. 3.3), the segment with genes 1 - 3 is combined with the segment with genes 4' - 10' to give a recombinant family. This is equivalent to the process of an equal crossing-over in terms of gene exchange and is treated separately by the equation (3.13), by replacing β with γ', for the rate of unequal crossing-overs. By this replacement, however, a correction term is needed that will be explained later. Note further that, in the actual process of inter-chromosomal unequal crossing-over, the deletion is the result of crossing-over, therefore the deletion phase also accompanies gene exchange. Then one cycle becomes again two unequal crossing-overs as in the case of intra-chromosomal one. In the following, I shall derive the mean, the variance and the covariance of the changes of gene lineage frequencies under the simple model.

Consider again the case where one cycle of an unequal crossing-over as in Fig. 3.3 has taken place between one of the i-th and one of the j-th family types. I calculate the mean, the variance and the covariance of the frequency change ($\Delta x_{i,k}$) by expansion and contraction process. Table 3.2 gives the necessary terms for the calculations. For comparison, see Table 2.1 which gives the corresponding calculations for the case of sister-chromatid crossing-over. At the expansion phase, $x_{i,k}$ increases by the amount $1/n_g$ with probability $x_{j,k}$. Similarly, $x_{i,\ell}$ increases with probability $x_{j,\ell}$. At the deletion phase, $x_{i,k}$ decreases by the amount $1/n_g$ with probability $x_{i,k}$, and so forth. Total changes of $x_{i,k}$ and $x_{i,\ell}$ in one cycle may be obtained by combining all cases and the last two columns show the results. Let us designate the mean change of $x_{i,k}$ by $M(\Delta x_{i,k})$, the variance of $\Delta x_{i,k}$ by $V(\Delta x_{i,k})$ and the covariance of $\Delta x_{i,k}$ and $\Delta x_{i,\ell}$ by $W(\Delta x_{i,k}\Delta x_{i,\ell})$. They are calculated by using the last two columns of Table 3.2 and are given at the bottom of the table.

It is necessary to calculate these quantities by taking one generation as a unit time. For that, one needs to consider all possible j as a partner of crossing-over. In other words, we take averages of these formulas by considering all family types in the population. We then multiply the formulas by the equivalent number of cycles of unequal crossing-overs, which is equal to m'γ'/2, since one crossing-over with shift of m' genes is equivalent to m'/2 cycles. The mean, the variance and the covariance of $\Delta x_{i,k}$ in one generation may be expressed by the following equations, by letting $\alpha' = m'\gamma'/n_g^2$.

Table 3.2 Derivation of the mean and variance of the change of gene lineage frequencies by inter-chromosomal unequal crossing-over.

Duplication phase			Deletion phase			Total		
Change of $x_{i,k}$	$x_{i,\ell}$	Probability	Change of $x_{i,k}$	$x_{i,\ell}$	Probability	Change of $x_{i,k}$	$x_{i,\ell}$	Probability
$+\frac{1}{n_g}$	0	$x_{j,k}$	$-\frac{1}{n_g}$	0	$x_{i,k}$	0	0	$x_{i,k}x_{j,k}$
			0	$-\frac{1}{n_g}$	$x_{i,\ell}$	$+\frac{1}{n_g}$	$-\frac{1}{n_g}$	$x_{i,\ell}x_{j,k}$
			0	0	$1-x_{i,k}-x_{i,\ell}$	$+\frac{1}{n_g}$	0	$x_{j,k}(1-x_{i,k}-x_{i,\ell})$
0	$+\frac{1}{n_g}$	$x_{j,\ell}$	$-\frac{1}{n_g}$	0	$x_{i,k}$	$-\frac{1}{n_g}$	$+\frac{1}{n_g}$	$x_{i,k}x_{j,\ell}$
			0	$-\frac{1}{n_g}$	$x_{i,\ell}$	0	0	$x_{i,\ell}x_{j,\ell}$
			0	0	$1-x_{i,k}-x_{i,\ell}$	0	$+\frac{1}{n_g}$	$x_{j,\ell}(1-x_{i,k}-x_{i,\ell})$
0	0	$1-x_{j,k}-x_{j,\ell}$	$-\frac{1}{n_g}$	0	$x_{i,k}$	$-\frac{1}{n_g}$	0	$x_{i,k}(1-x_{j,k}-x_{j,\ell})$
			0	$-\frac{1}{n_g}$	$x_{i,\ell}$	0	$-\frac{1}{n_g}$	$x_{i,\ell}(1-x_{j,k}-x_{j,\ell})$
			0	0	$1-x_{i,k}-x_{i,\ell}$	0	0	$(1-x_{i,k}-x_{i,\ell})(1-x_{j,k}-x_{j,\ell})$

$$M(\Delta x_{i,k}) = \frac{1}{n_g}(x_{j,k} - x_{i,k}), \quad V(\Delta x_{i,k}) = \frac{1}{n_g^2}\{x_{j,k}(1-x_{i,k}) + x_{i,k}(1-x_{j,k})\}$$

$$\text{and} \quad W(\Delta x_{i,k}, \Delta x_{i,\ell}) = \frac{-1}{n_g^2}\{x_{i,\ell}x_{j,k} + x_{i,k}x_{j,\ell}\}$$

$$M(\Delta x_{i,k}) = \frac{n_g \alpha'}{2} \underset{j}{E}\{x_{j,k} - x_{i,k}\} \tag{3.15}$$

$$V(\Delta x_{i,k}) = \frac{\alpha'}{2} \underset{j}{E}\{x_{j,k}(1 - x_{i,k}) + x_{i,k}(1 - x_{j,k})\} \tag{3.16}$$

and

$$W(\Delta x_{i,k} \; x_{i,\ell}) = \frac{-\alpha'}{2} \underset{j}{E}\{x_{i,\ell}x_{j,k} + x_{i,k}x_{j,\ell}\} \tag{3.17}$$

where $\underset{j}{E}$ denotes to taking the average for all j in the population.

By using the formulas (3.15) and (3.16), the changes of C_{w1} and C_{w2} in one generation become as follows,

$$
\begin{aligned}
E(\Delta C_{w1}) &= E\{\underset{k}{\Sigma}(x_{i,k} + \Delta x_{i,k})^2 - \Sigma \; x_{i,k}^2\} \\[2mm]
&= E\{\underset{k}{\Sigma}(2\Delta x_{i,k}x_{i,k} + \Delta x_{i,k}^2)\} \\[2mm]
&= E\left[\underset{k}{\Sigma}[n_g\alpha'\underset{j}{E}\{x_{j,k} - x_{i,k}\}x_{i,k} \right. \\[2mm]
&\qquad \left. + \frac{\alpha'}{2} \underset{j}{E}\{x_{j,k}(1 - x_{i,k}) + x_{i,k}(1 - x_{j,k})\}]\right] \\[2mm]
&= n_g\alpha'(C_{w2} - C_{w1}) + \alpha'(1 - C_{w2})
\end{aligned}
$$

and

$$
\begin{aligned}
E(\Delta C_{w2}) &= E\{\underset{k}{\Sigma}(x_{i,k} + \Delta x_{i,k})(x_{j,k} + \Delta x_{j,k}) - \underset{k}{\Sigma} \; x_{i,k}x_{j,k}\} \\[2mm]
&= E\{\underset{k}{\Sigma}(x_{i,k}\Delta x_{j,k} + x_{j,k}\Delta x_{i,k} + \Delta x_{i,k}\Delta x_{j,k})\} \\[2mm]
&= E\left[\underset{k}{\Sigma} \frac{n_g\alpha'}{2}\{x_{i,k} \underset{g}{E}(x_{g,k} - x_{j,k}) + x_{j,k} \underset{g}{E}(x_{g,k} - x_{i,k})\}\right] \\[2mm]
&= 0
\end{aligned}
$$

where E denotes to taking the expectation over the distribution of family types in the population. Thus, we have the following formula for the change due to unequal inter-chromosomal crossing-over,

$$C^{(V)} = \begin{bmatrix} 1 - n_g\alpha' & (n_g - 1)\alpha' \\ 0 & 0 \end{bmatrix} C^{(IV)} + \begin{bmatrix} \alpha' \\ 0 \end{bmatrix} . \qquad (3.19)$$

The equations (3.6), (3.7), (3.8), (3.13) and (3.19) give the total change of the identity coefficients in one generation by various forces. Note that α or α' is equivalent to the rate of approach to homozygosity due to finite population size in population genetics.

The above formulations are based on the assumption that the gene lineages are randomly arranged on the chromosome. This condition will be fully examined in the next section by using a new approach. Before going into detail of the new approach, we need a further clarification for the inter-chromosomal unequal crossing-over. As was stated previously, unequal inter-chromosomal crossing-over was treated partly by the expansion-contraction process and partly by the inter-chromosomal exchange process. For the latter one, the equation (3.13)(effect of equal crossing-over, page 26 - 27) is used by replacing β with γ'. Here it needs correction, since the effect of a certain fraction of the gene exchanges is already taken into account by the contraction-expansion process, as explained below, this fraction becomes particularly significant when m' is large. By referring to Fig. 3.2, the contribution of C_{w2} to C_{w1} is corrected as follows. When the mean shift at unequal crossing-over is by m' genes,

$$\gamma'[E\{2r(1 - r)\} - \mu'(1 - \mu')] \approx \frac{\gamma'}{3} \{1 - 3\mu'\} . \qquad (3.20)$$

where $\mu' = m'/n_g$. The correction term, $\mu'(1 - \mu') = 2\mu'(1 - \mu')/2$, comes from the consideration that, if one of the genes compared for identity in the recombinant family is one of the expanded gene members rather than that of the original set, such a comparison should be excluded in calculating the gene exchange rate, and the denominator 2 takes into account the deletion case. This is because such comparisons are already taken into account in calculating the effect of the contraction-expansion process. Incidentally, the correction term $\gamma'\mu'$ is equal to the coefficient $n_g\alpha' = \gamma'\mu'$ in the right hand side of equation (3.19) as it should be. The same reasoning applies to correcting the contribution of C_{w1} to C_{w2}. Thus, by using the equation (3.13), β should be replaced by $\gamma'\{1 - 3\mu'\}$ instead of γ'.

Now, it is possible to calculate the change of the identity coefficients in one generation, by combining the equations (3.6 - 8), (3.14) and (3.19), and by letting β' be the total corrected rate of inter-chromosomal crossing-over,

$$
C_{t+1} = (1 - 2v) \begin{bmatrix} 1 & 0 \\ \dfrac{1}{2N_e} & 1 - \dfrac{1}{2N_e} \end{bmatrix} \begin{bmatrix} 1 - \dfrac{\beta'}{3} & \dfrac{\beta'}{3} \\ 0 & 1 \end{bmatrix}
$$

$$
\times \begin{bmatrix} 1 - \alpha & 0 \\ 0 & 1 \end{bmatrix} \begin{bmatrix} 1 - n_g\alpha' & (n_g - 1)\alpha' \\ 0 & 1 \end{bmatrix} C_t + \begin{bmatrix} \alpha + \alpha' \\ 0 \end{bmatrix} \qquad (3.21)
$$

in which $\alpha = m\gamma/n_g^2$, $\alpha' = m'\gamma'/n_g^2$, $\beta' = \beta + \gamma'\{1 - 3\mu'\}$, and it is assumed that all parameters, v, $1/2N_e$, β', α and $n_g\alpha'$ are much less than unity. A most interesting question is; how much genetic variation is contained in a multigene family when equilibrium is reached among various forces? At the equilibrium, the identity coefficients remain unchanged. Therefore, $C_{t+1} = C_t$, and the solutions become

$$
\hat{C}_{w1} = \cfrac{\alpha + \alpha'}{2v + \dfrac{\beta'}{3}\dfrac{4N_e v}{1 + 4N_e v} + \alpha + \alpha'\dfrac{1 + 4n_g N_e v}{1 + 4N_e v}} \qquad (3.22)
$$

and

$$
\hat{C}_{w2} = \frac{\hat{C}_{w1}}{1 + 4N_e v} \ . \qquad (3.23)
$$

Note that the above formulas give the average identity coefficients at equilibrium which are equivalent to the identity coefficient of the two gene members randomly chosen from the family.

CHAPTER 4

DECREASE OF GENETIC CORRELATION WITH DISTANCE BETWEEN

GENE MEMBERS ON THE CHROMOSOME

The analyses in the previous chapter are based on several simplifying assumptions. In particular, the assumption of random arrangement of gene lineages on the chromosome is rather unrealistic. In the
present chapter, I shall present a more exact treatment of the problem
in the model of Kimura and Ohta (1979), with special reference to the
relationship between genetic correlation among gene members of a family
and their distance on the chromosome.

4.1 Unequal crossing-over by shift of one gene unit

As before, we consider a multigene family evolving under mutation,
random drift, equal and unequal crossing-over. In this section, let
us assume that intra-chromosomal unequal crossing-over occurs with
shift by one gene unit and inter-chromosomal crossing-over is always
equal. Later we shall consider more general cases. Also, we assume
that one gene is either duplicated or deleted each with probability
1/2 when one unequal crossing-over occurs. We use the same parameters
as before; γ = intra-chromosomal unequal crossing-over rate, β =
equal inter-chromosomal crossing-over rate and v = mutation rate.
At the moment we assume γ' (inter-chromosomal unequal crossing-over
rate) = 0 and m (mean number of gene shift at an unequal crossing-
over) = 1.

In the following, I derive the genetic correlation between gene
members of a family as a function of distance on the chromosome. We
shall denote by f_i the coefficient of identity between two genes
i-steps apart on one chromosome and by ϕ_i the identity coefficient
between two genes i-steps apart on different chromosomes (see Figure
4.1). Let us first formulate the case of no inter-chromosomal crossing over (β = 0). After one generation of unequal crossing-over and
mutation, f_i changes to f_i' according to the following formula (Kimura
and Ohta 1979);

$$f_i' \;=\; (1-v)^2 \left\{ (1 - \frac{\gamma i}{n_g}) f_i + \frac{\gamma i}{n_g} \frac{f_{i-1} + f_{i+1}}{2} \right\} \quad \text{for } i \geq 1 . \qquad (4.1)$$

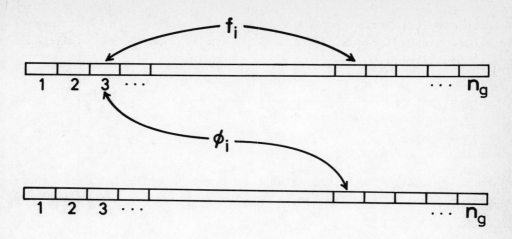

Figure 4.1   Diagram illustrating the meaning of
two identity coefficients f_i and ϕ_i
(from Kimura and Ohta 1979).

Note f_0 = 1 and $\gamma i/n_g$ is the probability that unequal crossing-over
occurs between two genes that are i-steps apart. Also note that f_i is
not influenced by random drift when β = 0, whereas ϕ_i changes by
mutation and random drift by the following formula. At the moment,
let us ignore the shift of positions by unequal crossing-overs in the
whole population for calculating ϕ_i. More exact analyses are in
progress.

$$\phi_i' \; = \; (1 - v)^2 \left\{ (1 - \frac{1}{2N_e}) \phi_i + \frac{1}{2N_e} f_i \right\} \; . \qquad (4.2)$$

Therefore, at statistical equilibrium in which f_i' = f_i and ϕ_i' =
ϕ_i, we obtain

$$2vf_i \; = \; \frac{\gamma i}{n_g} \frac{f_{i+1} - 2f_i + f_{i-1}}{2} \qquad (4.3)$$

and

$$\phi_i \; = \; \frac{f_i}{1 + 4N_e v} \qquad (4.4)$$

where we neglect terms involving $v/2N_e$ and v^2. When n_g is sufficiently large, the difference equation (4.3) becomes the following differential equation;

$$2vf(x) = \gamma x \frac{f(x+\frac{1}{n_g}) - 2f(x) + f(x-\frac{1}{n_g})}{2} \approx \frac{\gamma x}{2n_g^2} \frac{d^2f(x)}{dx^2} \qquad (4.5)$$

$$\text{for} \quad 0 \leq x \leq 1$$

where $x = i/n_g$. The solution of this equation which satisfies the condition $f(0) = 1$ and which is bounded as $a \to \infty$ is (Kimura and Ohta 1979);

$$f(x) = 2\sqrt{ax} \, K_1(2\sqrt{ax}) \qquad (4.6)$$

where $a = 4n_g^2 v/\gamma$ and $K_1(\cdot)$ denotes a Modified Bessel function.

When inter-chromosomal equal crossing-over takes place ($\beta > 0$), the formulation is more complicated, however Kimura and Ohta (1979) obtained the solution at equilibrium. First, by intra-chromosomal unequal crossing-over, f_i and ϕ_i change by the following equation,

$$f_i' = (1 - \frac{i\gamma}{n_g})f_i + \frac{i\gamma}{n_g} \frac{f_{i-1} + f_{i+1}}{2} \qquad \text{for } i \geq 1$$

$$\qquad (4.7)$$

$$\phi_i' = \phi_i$$

Next, by inter-chromosomal equal crossing-over with rate β per generation,

$$f_i'' = (1 - \frac{i\beta}{n_g})f_i' + \frac{i\beta}{n_g} \phi_i'$$

$$\qquad (4.8)$$

$$\phi_i'' = (1 - \frac{1}{2N_e-1} \cdot \frac{i\beta}{n_g})\phi_i' + \frac{1}{2N_e-1} \cdot \frac{i\beta}{n_g} f_i'$$

in which the coefficient $\frac{1}{2N_e-1} \cdot \frac{i\beta}{n_g}$ in the second equation comes from the same reasoning as the case of equation (3.13) in page 27. In other words, after inter-chromosomal crossing-overs, the two genes that are i steps apart and that are located on one chromosome become separated into two different chromosomes with probability $i\beta/n_g$, and there are $2N_e - 1$ ways of matching the first gene on this chromosome with the gene at the second site on any other chromosome. By random genetic drift and mutation, f_i'' and ϕ_i'' change to f_i''' and ϕ_i''' by the following equation.

$$f_i''' = (1 - v)^2 f_i''$$

$$\phi_i''' = (1 - v)^2 \left\{ (1 - \frac{1}{2N_e}) \phi_i'' + \frac{1}{2N_e} f_i'' \right\} .$$

(4.9)

Total change of f_i and ϕ_i in one generation becomes by combining the equations (4.7 - 4.9);

$$f_i''' = (1 - v)^2 \left\{ (1 - \frac{i\beta}{n_g}) [f_i + \frac{i\gamma}{2n_g}(f_{i-1} - 2f_i + f_{i+1})] + \frac{i\beta}{n_g} \phi_i \right\}$$

(4.10)

$$\phi_i''' = (1 - v)^2 \left\{ \left[(1 - \frac{1}{2N_e}) \frac{1}{2N_e-1} \cdot \frac{i\beta}{n_g} + \frac{1}{2N_e}(1 - \frac{i\beta}{n_g}) \right] \right.$$

$$\times \left[f_i + \frac{i\gamma}{2n_g}(f_{i-1} - 2f_i + f_{i+1}) \right]$$

$$\left. + \left[(1 - \frac{1}{2N_e})(1 - \frac{1}{2N_e-1}\frac{i\beta}{n_g}) + \frac{1}{2N_e}\frac{i\beta}{n_g} \right] \phi_i \right\} .$$

(4.11)

When n_g is sufficiently large, the above equations again reduce to differential equations, which take the following form at equilibrium, after neglecting small terms such as $v/2N_e$ or v^2.

$$\frac{\gamma}{2n_g^2}(1 - \beta x) x \frac{d^2}{dx^2} f(x) - (2v + \beta x) f(x) + \beta x \phi(x) = 0$$

(4.12)

and

$$\frac{\gamma}{2n_g^2} x \frac{d^2}{dx^2} f(x) + f(x) - (1 + 4N_e v) \phi(x) = 0$$

(4.13)

in which $x = i/n_g$. From these equations we again get,

$$\phi(x) \approx \frac{f(x)}{1 + 4N_e v} .$$

(4.14)

Also, the differential equation for f(x) is obtained,

$$x \frac{d^2}{dx^2} f(x) - a(1+ bx)f(x) = 0, \qquad (x \le x \le 1), \qquad (4.15)$$

where $a = 4n_g^2 v/\gamma$ and $b = 2N_e\beta/(1 + 4N_e v)$ and we assume $\beta \ll 1$. Equation (4.15) is Whittaker's differential equation (Whittaker and Watson 1935) and Kimura and Ohta (1979) obtained the pertinent solution, which reduces to (4.6) at the limit $b \to 0$, in the following form

$$f(x) = \Gamma(1 - K)W_{K,1/2}(2\sqrt{ab}\ x), \qquad (4.16)$$

where $K = -(1/2)\sqrt{a/b}$ and $W_{K,1/2}(z)$ is Whittaker's function. For numerical evaluation, it is more convenient to express this solution in the following form.

$$f(x) = e^{-\sqrt{ab}\ x} (2\sqrt{ab}\ x) \int_0^\infty e^{-2\sqrt{ab}\ xt}(\frac{t}{1+t})^{\frac{1}{2}\sqrt{\frac{a}{b}}} dt. \quad (4.17)$$

Figure 4.2 gives some examples of f(x) calculated by equation (4.6).

4.2 Unequal crossing-over by shift of more than one gene

Let us consider a more general situation in which the shift at an unequal crossing-over is not restricted to one gene, and also unequal inter-chromosomal crossing-over may occur. Let j be the number of genes of shift at intra-chromosomal unequal crossing-over. We shall assume that j follows the probability distribution,

$$p_j = \frac{2}{\ell} (1 - \frac{j - 0.5}{\ell}) \qquad \text{for } j \le \ell$$

$$(4.18)$$

$$p_j = 0 \qquad \text{for } j > \ell$$

in which ℓ is the maximum gene number of shift. This function implies that the smaller the shift is, the more likely it occurs. It is considered to be in accord with the previous simulation studies (Smith 1974, Ohta 1978a). We also assume that, with probability 1/2, j genes are either duplicated or deleted by an unequal

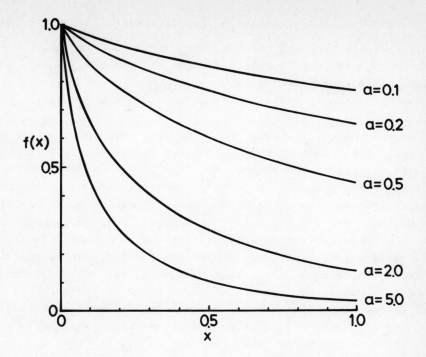

Figure 4.2   Relationship between identity coefficient,
             f(x), and the distance, x, on the chromo-
             some when b = 0 (from Kimura and Ohta 1979).

crossing-over. The parameter β is the rate of equal inter-chromosomal
crossing-over, and at the moment, we assume that all inter-chromosomal
crossing-over are equal.

It is possible to formulate the transition of f_i and ϕ_i from one
generation to the next, by using the same method as used in the
previous section. After one generation of unequal crossing-over and
mutation, f_i changes to f_i' according to the formula;

$$f_i' = (1 - v)^2 \left[\left(1 - \frac{i\gamma}{n_g}\right) f_i + \frac{i\gamma}{2n_g} \left\{ \sum_{j=1}^{\ell} (f_{|i-j|} + f_{i+j}) p_j \right\} \right] \qquad (4.19)$$

where $i \geq 1$, and $|i - j|$ is the absolute value of $i - j$. Note that

$f_0 \equiv 1$ and $\sum_{j=1}^{\ell} p_j = 1$. In the above formulation, $\gamma i/n_g$ is the probability that an unequal crossing-over occurs between the two genes that are i-steps apart. In Figure 4.3, an example of unequal crossing-over is depicted whereby genes that are $|i-j|$ and $(i + j)$ steps apart are each converted to have distance i in the next generation.

As to the change of ϕ_i, the shift of positions in the whole population needs to be considered. When, after an unequal crossing-over, j genes are shifted in the recombinant family, all pairwise comparisons involving the recombinant family for calculating ϕ_i in the population are affected. On the average, one half of the positions of such comparisons are shifted by j gene units. When the comparisons are between the recombinant family and another recombinant family, however, the shift of positions is much complicated, yet nearly zero on the average, and such comparisons are subtracted in calculating the change of ϕ_i. Thus, the proportion, $2\gamma(1 - \gamma)/2$ of ϕ_i would be affected in one generation through unequal intra-chromosomal crossing-overs. Thus ϕ_i changes by mutation and intra-chromosomal crossing-over to ϕ_i' according to the following relation,

$$\phi_i' = (1 - v)^2 \left[\{1 - \gamma(1-\gamma)\}\phi_i + \frac{\gamma(1-\gamma)}{2} \{\sum_j (\phi_{|i-j|} + \phi_{i+j}) p_j\} \right].$$

$$(4.20)$$

To be exact, a similar term should be included in equations (4.2) and (4.7) when the shift occurs always by one gene as in the previous section. Exact analyses for the cases with shift only by one gene unit are under progress.

The change of f_i and ϕ_i due to inter-chromosomal equal crossing-over and random drift are the same as in the previous section. Through one generation of inter-chromosomal equal crossing-overs (with rate β) and random drift, f_i and ϕ_i change according to the equation,

$$f_i'' = (1 - \frac{i\beta}{n_g}) f_i' + \frac{i\beta}{n_g} \phi_i'$$

$$(4.21)$$

$$\phi_i'' = (1 - \frac{1}{2N_e})(1 - \frac{1}{2N_e-1} \cdot \frac{i\beta}{n_g})\phi_i' + \frac{i\beta}{2N_e n_g} \phi_i'$$

$$+ \left\{ (1 - \frac{1}{2N_e}) \frac{1}{2N_e-1} \cdot \frac{i\beta}{n_g} + \frac{1}{2N_e}(1 - \frac{i\beta}{n_g}) \right\} f_i'$$

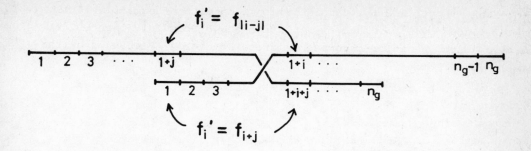

Figure 4.3   Diagram showing the change of identity
coefficient by an unequal crossing-over
with shift of j genes.

By using equations (4.19) - (4.21), it is possible to generate numerically f_i and ϕ_i from one generation to the next. The most interesting
situation is the equilibrium when various forces balance each other.
In the next section, the equilibrium values of f_i and ϕ_i will be
obtained for some interesting cases and will be compared with the approximate result (equation 3.22).

Next, we shall consider the case where inter-chromosomal crossing-
over may contain an unequal one. In addition to the equal inter-
chromosomal crossing-over with the rate β, we shall assume that unequal inter-chromosomal crossing-over occurs with the rate γ' per
family per generation. Let us further assume that the number of gene
shift, j, at unequal crossing-over follows the same kind of probability function as before,

$$q_j = \frac{2}{\ell'}(1 - \frac{j - 0.5}{\ell'}) \qquad \text{for } j \leq \ell'$$

$$q_j = 0 \qquad \text{for } j > \ell' \tag{4.22}$$

in which ℓ' is the maximum number of gene shift. Then it is possible
to formulate the change of f_i and ϕ_i in one generation due to inter-
chromosomal unequal crossing-over.

$$f_i''' = (1 - \frac{i\gamma'}{n_g}) f_i'' + \frac{i\gamma'}{2n_g} \left\{ \sum_{j=1}^{\ell'} (\phi''_{|i-j|} + \phi''_{i+j}) q_j \right\} \qquad \text{for } i \geq 1. \tag{4.23}$$

Note that $f_{|i-j|}$ and f_{i+j} in the right hand side of equation (4.19) are replaced by $\phi''_{|i-j|}$ and ϕ''_{i+j} in the above formula, since we are dealing with the inter-chromosomal crossing-over. ϕ_i changes by the following equation.

$$\phi_i''' = \left\{ 1 - \frac{1}{2N_e - 1} \cdot \frac{i\gamma'}{n_g} - \gamma'(1-\gamma') \right\}\phi_i''$$

$$+ \frac{1}{2N_e - 1} \cdot \frac{i\gamma'}{2n_g} \left\{ \sum_{j=1}^{\ell'} (f''_{|i-j|} + f''_{i+j})q_j \right\}$$

$$+ \frac{\gamma'(1-\gamma')}{2} \left\{ \sum_{j=1}^{\ell'} (\phi''_{|i-j|} + \phi''_{i+j})q_j \right\} \qquad \text{for } i \geq 0. \qquad (4.24)$$

Note $f_0 = 0$ but $\phi_0 \neq 0$, and ϕ_0 comes into equation (4.24). The last term in the above formula is the effect of shift of gene positions in the population as a whole just as in the case of intra-chromosomal unequal crossing-over (equation 4.20). Also, the coefficient, $(1/(2N_e - 1))(i\gamma'/n_g)$, comes from the same reasoning as used for treating inter-chromosomal equal crossing-overs (equation 3.13). By using equations (4.23) and (4.24), it is possible to calculate the change of identity coefficients by unequal inter-chromosomal crossing-overs. The results of numerical calculations are given below.

Equilibrium identity coefficients are numerically calculated. Starting from the homogeneous population ($f_i = \phi_i = 1.0$ at $t = 0$), f_i and ϕ_i are generated from one generation to the next by using equations (4.19) - (4.21), (4.23) and (4.24). The equilibrium values are obtained at the $(2/v)$-th generation. In general, there is a discontinuity between f_0 and f_1, since $f_0 \equiv 1$ but f_i may be considerably less than unity when the shift may occur by more than one gene. Then the relationship between f_i and ϕ_i appears to be more complicated than the previous case (equation 4.4). In particular, ϕ_0 may be much less than $f_0 = 1$ and close to ϕ_1 due to the shift of positions.

The identity coefficient (f_i) is given as a function of chromosomal distance, i/n_g. Figure 4.4 represents the case of intra-chromosomal unequal crossing-over ($\gamma = 0.5$, $\gamma' = 0$). Also, Figure 4.5 represents the case of inter-chromosomal unequal crossing-over ($\gamma' = 0.5$, $\gamma = 0$) and the case of both intra- and inter-chromosomal ones ($\gamma = \gamma' = 0.25$). Other parameters are; $v = 10^{-3}$, $\beta = 0.01$ and $n_g = 50$ (Fig. 4.4) and $v = 10^{-4}$, $\beta = 0.0$ and $n_g = 50$ (Fig. 4.5). Note a discontinuity between f_0 and f_1, where $f_0 \equiv 1$. In both figures, it may be seen that, for larger ℓ or ℓ', curves become more flat, indicating that the identity coefficient becomes more independent of chromosomal distance between the two genes.

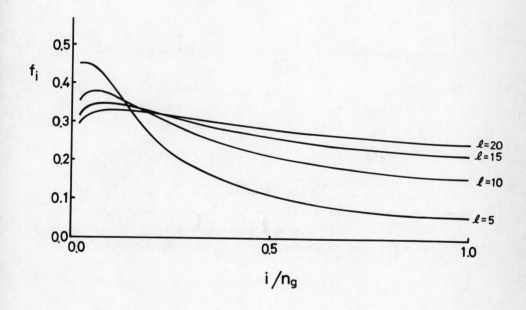

Figure 4.4. Relationship between identity coefficient,
f_i, and the distance, i/n_g, on the chromosome, when intra-chromosomal crossing-over
occurs with shift of more than one gene.

Parameters are; $\gamma = 0.5$, $\gamma' = 0$, $v = 10^{-3}$,
$\beta = 10^{-2}$, $n_g = 50$ and $2N_e = 100$.

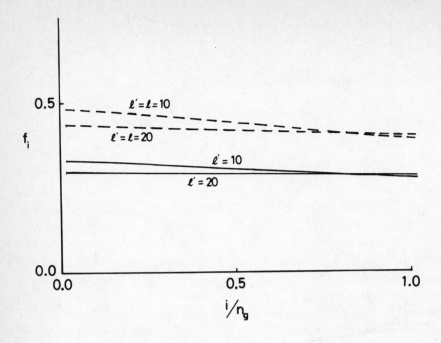

Figure 4.5   Relationship between identity coefficient,
f_i, and the chromosomal distance, i/n_g, when
inter- and intra-chromosomal crossing-overs
occur with shift of more than one gene.
Parameters are; $v = 10^{-4}$, $\beta = 0.0$, $n_g = 50$,
$2N_e = 100$ for all curves, and $\gamma = \gamma' = 0.25$
for the broken curves and $\gamma = 0$ and $\gamma' = 0.5$
for the solid curves.

4.3   Average identity coefficient

Based on the theory developed here, let us calculate the average
identity coefficient which is the probability of two randomly chosen
genes from the multigene family being identical.   The average values
may be compared with the approximate result obtained in chapter 3.
The average identity coefficient may be calculated by the following
formula (Kimura and Ohta 1979).

$$\bar{f} = \hat{c}_{W1} = \frac{2}{n_g(n_g+1)} \left\{ n_g f_0 + (n_g-1) f_1 + \cdots + f_{n_g-1} \right\} . \qquad (4.25)$$

When n_g is large, and the analytical expression for f_i is available, this formula may be replaced by

$$\bar{f} = 2 \int_0^1 (1 - x) f(x) \, dx \qquad (4.26)$$

where $x = i/n_g$. Equation (3.22), which is an approximate equation, is based on the assumption that individual gene lineages are randomly arranged on the chromosome, so that f_i is constant. Thus, when f_i is independent of chromosomal distance as in cases with large ℓ or ℓ' of the examples in the previous section, the approximate results should give correct values, whereas when ℓ or ℓ' is small, they are likely to be biased. However, when the shift of gene number at an unequal crossing-over is exactly one as in the model of section 4.1, the agreement is expected to be better. This is because the diffusion of gene lineage in the chromosome is appropriately treated when the shift is restricted to one gene unit even when genetic correlation with chromosomal distance is not negligible.

An approximate solution (3.22) is

$$\bar{f} = \hat{c}_{W1} = \frac{\alpha + \alpha'}{2v + \dfrac{\beta'}{3}\dfrac{4N_e v}{1+4N_e v} + \alpha + \alpha'\dfrac{1+4n_g N_e v}{1+4N_e v}} \qquad (4.27)$$

in which $\alpha = m\gamma/n_g^2$ and $\alpha' = m'\gamma'/n_g^2$ with m and m' respectively representing the average number of gene shifts at intra- and inter-chromosomal unequal crossing-over. The parameter β' is the total rate of inter-chromosomal crossing-over,

$$\beta' = \beta + \gamma'(1 - 3\mu') \qquad (4.28)$$

in which $\mu' = m'/n_g$ and the term $3\mu'$ is a correction factor (see page 32). In the present model of shift at unequal crossing-over (equation 4.22), the mean gene number of shift is

$$m = \frac{(\ell + 1)(2\ell + 1)}{6\ell} \quad , \quad \text{and} \quad m' = \frac{(\ell' + 1)(2\ell' + 1)}{6\ell'} \quad (4.29)$$

In the simple model of a shift of one gene unit at an intra-chromosomal unequal crossing-over (section 4.1), we have m = 1 and m' = 0.

Table 4.1 gives a comparison between the exact and the approximate results for the simple case of intra-chromosomal unequal crossing-over of one gene shift with no inter-chromosomal crossing-over (m = 1, β = γ' = 0). The exact values are obtained by using equations (4.6) and (4.26). As can be seen from the table, the agreement is satisfactory, although the approximate formula seems to give a slight overestimation.

Table 4.1   Average identity coefficient ($\hat{C}_{w1}$) for various values of a, assuming no inter-chromosomal crossing-over.

a	Eqs. (4.6) and (4.26) (exact)	Equation (4.27) (approximate)
0.05	0.9397	0.9756
0.1	0.8993	0.9524
0.2	0.8378	0.9091
0.3	0.7898	0.8696
0.4	0.7498	0.8333
0.5	0.7155	0.8000
0.8	0.6340	0.7143
1.0	0.5916	0.6667
2.0	0.4505	0.5000
5.0	0.2698	0.2857
8.0	0.1933	0.2000

(a = $4n_g^2 v/\gamma$)

Next, let us examine the case where an unequal crossing-over occurs by shift of more than one gene unit (m > 1 or m' > 1). The equilibrium values are obtained by repeating calculations over 2/v generations starting from a homogeneous gene family at the start, i.e. $f_i = \phi_i = 1$. The average identity coefficients are computed by using equation (4.25). A comparison of the results obtained by using approximate formula (4.27) and the exact numerical results by using equations (4.19) - (4.24) and (4.25) is given in Table 4.2. Broadly speaking, the agreement is satisfactory when m or m' is more than 10% of the average family size, n_g. When m or m' is smaller, the approximate formula underestimates the true value. This is because the diffusion of gene lineage by unequal crossing-overs is underestimated when the genetic correlation decreases with the chromosomal distance. This conclusion is consistent with the empirical results of the Monte Carlo experiments (Smith 1974, Ohta 1978a). Also, under the approximate treatment, the effect of inter-chromosomal exchange is exaggerated when genetic correlation decreases with chromosomal distance. In other words, when m or m' is less than 10% of n_g, the assumption of random arrangement of gene lineages does not hold, and therefore the effect of gene exchange between the families is overestimated under the assumption of random arrangement of gene lineages. The bias becomes larger as β and/or γ' gets larger. Since both the overestimation of gene exchange and underestimation of gene lineage diffusion result in decrease of identity coefficient, the disagreement between the approximate and the exact results is fairly large when m or m' is small and β or γ' is large. However, when m or m' is larger than 10% of n_g, the two results are in good agreement.

I have performed numerical calculations for various values of n_g other than 50 and obtained similar results, indicating that the average shift of 10% is a borderline beyond which significant genetic correlations as the function of chromosomal distances arise. Although we have confined our analyses to the case of one specific distribution for the shift at unequal crossing-overs (equations 4.18 and 4.22), the method may be extended for certain different forms of probability function. Furthermore the theory may be applicable to the model of gene conversion if a probability function analogous to the formula (4.18) or (4.22) is known for gene correction. The results in this chapter give the theoretical basis to judge the adequacy and limitation of the approximate analyses assuming double diffusion (chapters 3 and 5).

Table 4.2 Average identity coefficient when the shift at
unequal crossing-over is more than one gene
unit. Letters ℓ and ℓ' stand for the maximum
of gene shift at intra- and inter-chromosomal
unequal crossing-overs.

Parameters	ℓ or ℓ'	Exact $\hat{C}_{wl}$	Approximate $\hat{C}_{wl}$
$v = 5 \times 10^{-4}$	5	0.4257	0.2892
$\beta = 10^{-2}$	10	0.4935	0.4159
$2N_e = 25$	15	0.5139	0.5048
$\gamma = 0.5$	20	0.5193	0.5703
$\gamma' = 0$			
$v = 10^{-3}$	5	0.2543	0.1469
$\beta = 10^{-2}$	10	0.3005	0.2315
$2N_e = 100$	15	0.3170	0.3013
$\gamma = 0.5$	20	0.3218	0.3596
$\gamma' = 0$			
$v = 10^{-4}$	10	0.4591	0.2966
$\beta = 0$	15	0.4531	0.3768
$2N_e = 100$	20	0.4439	0.4409
$\gamma = \gamma' = 0.25$			
$v = 10^{-4}$	10	0.3368	0.1823
$\beta = 0$	15	0.3292	0.2423
$2N_e = 100$	20	0.3216	0.2944
$\gamma = 0$			
$\gamma' = 0.5$			

CHAPTER 5

GENETIC DIFFERENTIATION OF MULTIGENE FAMILIES

5.1 Identity coefficient between the populations

Let us consider a situation where a population was split into two identical ones and the two populations have been isolated thereafter. Let N_e be the effective population size of both new and original populations, i. e. the original population has doubled to give the two daughter populations. I shall here apply the simple model of double diffusion. Let us assume that the population and gene family size remain unchanged for a long period, so that genetic variations contained in the families of one population fluctuate only with the statistically predictable latitude, whereas the gene families differentiate with time between the two populations.

The situation may be illustrated by the following diagram (Fig. 5.1). As shown in the figure, $x_{i,k}$ and $y_{\ell,k}$ are the frequencies of the k-th gene lineage in the i-th family type of the first population and in the ℓ-th family type of the second population respectively. The identity coefficient may be measured between the two populations in addition to the two measures (for one chromosome and for different chromosomes) within a population investigated previously. In total, one can define the following three measures of identity:

Single chromosome
$$c_i = \sum_k x_{i,k}^2 \qquad (5.1)$$

Two chromosomes from the same population
$$c_{ij} = \sum_k x_{i,k} x_{j,k} \qquad (5.2)$$

Two chromosomes each taken from different populations
$$e_{i\ell} = \sum_k x_{i,k} y_{\ell,k} \cdot \qquad (5.3)$$

where summation is over all gene lineages. Note that $e_{i\ell}$ is the identity coefficient between the i-th family type in the first population and the ℓ-th family type in the second population.

The same parameters are used as before; v = mutation rate, γ = intra-chromosomal unequal crossing-over rate with mean shift of m genes, β = inter-chromosomal equal crossing-over rate, γ' = inter-chromosomal unequal crossing-over rate with the mean shift of m' genes, and n_g is the mean gene family size.

The average values of these identity coefficients over all family types in the two populations may be expressed by the following formulas, by referring to Fig. 5.1,

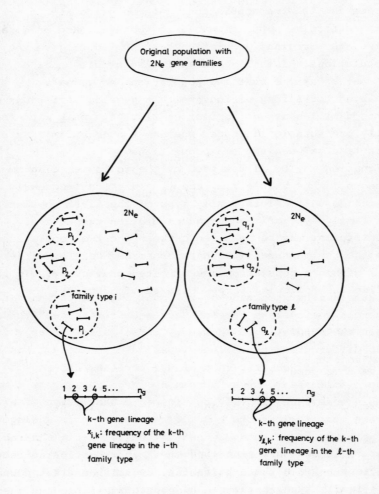

Figure 5.1 Diagram showing a multigene family
            in two isolated populations.

$$C_{w1} = \sum_i c_i p_i \tag{5.4}$$

$$C_{w2} = \sum_i \sum_j c_{ij} p_i p_j \tag{5.5}$$

for the first population. The average identity coefficient between the populations becomes a function of the time since divergence, t.

$$C_b(t) = \sum_i \sum_\ell e_{i\ell} p_i q_\ell(t) \tag{5.6}$$

where the summation is for all family types in the population. At equilibrium, $E\{C_{w1}\} = \hat{C}_{w1}$ and $E\{C_{w2}\} = \hat{C}_{w2}$ which are given by the equations (3.22) and (3.23) under the simple model and take the same value both in the first and in the second populations.

On the other hand, the expected value of $C_b(t)$ decreases with time by the constant rate, equal to twice the mutation rate, starting from $\hat{C}_{w2}$.

$$\hat{C}_b(t) = E\{C_b(t)\} = E\{C_b(0)\}e^{-2vt} = \hat{C}_{w2}e^{-2vt} . \tag{5.7}$$

This is because neither crossing-over (equal and unequal) nor sampling drift influences the average rate of decrease of the identity coefficient between the populations. However, these forces greatly influence genetic differentiation of individual cases. In fact, in spite of decreasing identity coefficients at the constant rate on the average, the differentiation may accompany large variance; it may be quite rapid in some cases and may be slow in others, even if all parameters are more or less the same. The situation is analogous to the differentiation of selectively neutral enzyme loci (Chakraborty et al. 1978), although our problem is much more complicated because of the doubly stochastic process (unequal crossing-overs and random genetic drift). It is necessary to understand this process theoretically. As will be discussed later, there are some interesting examples to which the results may be applied.

5.2   Variance and covariance of identity coefficients

In this section, I shall derive the variance and covariance of
the identity coefficients within and between populations. The diffusion method is used, since it provides a simple way of calculating the
rate of changes of moments of random variables. As shown in Fig. 5.1,
both $x_{i,k}$ (frequency of the k-th gene lineage in the i-th gene family
type) and p_i (frequency of the i-th gene family type in the first
population) are random variables and identity coefficients are expressed as their moments (formulas 5.1 - 5.3). The rate of change of
moments may be obtained by using the following diffusion equation
(Ohta and Kimura 1971)

$$\frac{d}{dt} E\{f(r_1, r_2, \cdots r_n)\} = E[L\{f(r_1, r_2, \cdots r_n)\}] \qquad (5.8)$$

where r_i is the i-th random variable, f is a continuous function of
r_i's and L is the differential operator of the Kolmogorov backward
equation in the following form

$$L = \frac{1}{2} \sum_i V_{\delta r_i} \frac{\partial^2}{\partial r_i^2} + \sum_{i>j} W_{\delta r_i \delta r_j} \frac{\partial^2}{\partial r_i \partial r_j} + \sum_i M_{\delta r_i} \frac{\partial}{\partial r_i} \qquad (5.9)$$

in which M, V and W designate, respectively, the mean, the variance
and the covariance of the rates of changes of the r_i's that appear as
subscripts. In our case, the random variables are $x_{i,k}$ and $y_{\ell,k}$ for
treating unequal crossing-overs and p_i and q_ℓ for treating random
genetic drift. M, V and W for $x_{i,k}$ for intra- and inter-chromosomal
unequal crossing-overs were obtained in chapter 3. By taking one
generation as unit time, they may be expressed as follows. For the
first population,

$$M(\Delta x_{i,k}) = \frac{n_g \alpha'}{2} \sum_j E\{x_{j,k} - x_{i,k}\} \qquad (5.10)$$

$$V(\Delta x_{i,k}) = \alpha x_{i,k}(1 - x_{i,k})$$

$$+ \frac{\alpha'}{2} \sum_j E\{x_{j,k}(1 - x_{i,k}) + x_{i,k}(1 - x_{j,k})\} \qquad (5.11)$$

$$W(\Delta x_{i,k},\ \Delta x_{i,g}) = -\alpha x_{i,k} x_{i,g}$$

$$- \frac{\alpha'}{2} \sum_j E\{x_{i,g} x_{j,k} + x_{i,k} x_{j,g}\} \tag{5.12}$$

where $\alpha = m\gamma/n_g^2$ and $\alpha' = m'\gamma'/n_g^2$. Similarly for the second population,

$$M(\Delta y_{\ell,k}) = \frac{n_g \alpha'}{2} \sum_r E\{y_{r,k} - y_{\ell,k}\} \tag{5.13}$$

$$V(\Delta y_{\ell,k}) = \alpha y_{\ell,k}(1 - y_{\ell,k})$$

$$+ \frac{\alpha'}{2} \sum_r E\{y_{r,k}(1 - y_{\ell,k}) + y_{\ell,k}(1 - y_{r,k})\} \tag{5.14}$$

and

$$W(\Delta y_{\ell,k},\ \Delta y_{\ell,g}) = -\alpha y_{\ell,k} y_{\ell,g}$$

$$- \frac{\alpha'}{2} \sum_r E\{y_{\ell,g} y_{r,k} + y_{\ell,k} y_{r,g}\} \ . \tag{5.15}$$

It is also clear that the covariance of $\Delta x_{i,k}$ and $\Delta y_{\ell,k}$ is zero.

$$W(\Delta x_{i,k},\ \Delta y_{\ell,k}) = 0 \tag{5.16}$$

Now, in order to calculate the changes of moments due to random genetic drift, the well known formulas for M, V and W for Δp_i and Δq_ℓ may be expressed by the following equations, again by taking one generation as unit time (e. g. Crow and Kimura 1970, p. 330).

$$M(\Delta p_i) = 0 \tag{5.17}$$

$$V(\Delta p_i) = p_i(1 - p_i)/2N_e \tag{5.18}$$

and

$$W(\Delta p_i \Delta p_j) = -p_i p_j/2N_e \tag{5.19}$$

for the first population. The corresponding quantities for the second population may be obtained by replacing p_i with q_ℓ and p_j by q_m. Also $W(\Delta p_i,\ \Delta q_\ell) = 0$.

Now, it is possible to calculate the rate of changes of moments, due to unequal crossing-over and random genetic drift, by using the above equations (5.8) - (5.19).

In order to obtain the variance and covariance of identity coefficients, one needs the following set of functions, by letting again E denote taking expectation for family type distributions in the two populations. One population, third moments:

$$D_{w1} = E\{\sum_k x_{i,k}^3\} = \sum_i d_i p_i \qquad (5.20)$$

$$D_{w2} = E\{\sum_k x_{i,k}^2 x_{j,k}\} = \sum_i \sum_j d_{ij} p_i p_j \qquad (5.21)$$

$$D_{w3} = E\{\sum_k x_{i,k} x_{j,k} x_{g,k}\} = \sum_i \sum_j \sum_g d_{ijg} p_i p_j p_g \qquad (5.22)$$

where $d_i = \sum_k x_{i,k}^3$, $d_{ij} = \sum_k x_{i,k}^2 x_{j,k}$ and $d_{ijg} = \sum_k x_{i,k} x_{j,k} x_{g,k}$.

Fourth moments:

$$E_{w11} = E(c_i^2) = E\{(\sum_k x_{i,k}^2)^2\} = \sum_i c_i^2 p_i \qquad (5.23)$$

$$E_{w12} = \frac{1}{2} E\{(c_i + c_j)c_{ij}\} = E\{(\sum_k x_{i,k}^2)(\sum_k x_{i,k} x_{j,k})\}$$

$$= \frac{1}{2} \sum_i \sum_j (c_i + c_j) c_{ij} p_i p_j \qquad (5.24)$$

$$E_{w22} = E(c_{ij}^2) = E\{(\sum_k x_{i,k} x_{j,k})^2\} = \sum_i \sum_j c_{ij}^2 p_i p_j \qquad (5.25)$$

$$E_{w3} = E(c_{ij} c_{ig}) = E\{(\sum_k x_{i,k} x_{j,k})(\sum_k x_{i,k} x_{g,k})\}$$

$$= \sum_i (\sum_j \sum_g c_{ij} c_{ig} p_j p_g) p_i \qquad (5.26)$$

$$F_{w11} = E(C_{w1}^2) = E\{(\sum_i c_i p_i)^2\} \qquad (5.27)$$

$$F_{w12} = E(C_{w1} C_{w2}) = E\{(\sum_i c_i p_i)(\sum_i \sum_j c_{ij} p_i p_j)\} \qquad (5.28)$$

$$F_{w22} = E(C_{w2}^2) = E\{(\sum_i \sum_j c_{ij}p_ip_j)^2\} \tag{5.29}$$

$$F_{w2} = E\{\overline{c_{ij/i}}^2\} = \sum_i(\sum_j c_{ij}p_j)^2 p_i \tag{5.30}$$

where $\overline{c_{ij/i}}$ is the mean of c_{ij} for a given i.

The moments concerning the identity probability between the two populations are functions of time since divergence (t), however in the following I shall omit t for simplicity. Third moments:

$$D_{b2} = E\{\sum_k x_{i,k}^2 y_{\ell,k}\} = \sum_i \sum_j f_{i\ell}p_iq_\ell \tag{5.31}$$

$$D_{b3} = E\{\sum_k x_{i,k}x_{j,k}y_{\ell,k}\} = \sum_i \sum_j \sum_\ell f_{ij\ell}p_ip_jq_\ell \tag{5.32}$$

where $f_{i\ell} = \sum_k x_{i,k}^2 y_{\ell,k}$ and $f_{ij\ell} = \sum_k x_{i,k}x_{j,k}y_{\ell,k}$. Note that $x_{i,k}$ and p_i are random variables concerning the first population, whereas $y_{\ell,k}$ and q_ℓ are those concerning the second population. Fourth moments:

$$E_{w1b} = E\{c_ie_{i\ell}\} = E\{\sum_k x_{i,k}^2 \sum_k x_{i,k}y_{\ell,k}\}$$
$$= \sum_i \sum_\ell c_ie_{i\ell}p_iq_\ell \tag{5.33}$$

$$E_{w2b} = E\{c_{ij}e_{i\ell}\} = E\{\sum_k x_{i,k}x_{j,k} \sum_k x_{i,k}y_{\ell,k}\}$$
$$= \sum_i \sum_j \sum_\ell c_{ij}e_{i\ell}p_ip_jq_\ell \tag{5.34}$$

$$E_{b22} = E\{e_{i\ell}^2\} = E\{(\sum_k x_{i,k}y_{\ell,k})^2\} = \sum_i \sum_\ell e_{i\ell}^2 p_iq_\ell \tag{5.35}$$

$$E_{b3} = E\{e_{ig}e_{i\ell}\} = E\{\sum_k x_{i,k}y_{g,k} \sum_k x_{i,k}y_{\ell,k}\}$$
$$= \sum_i \sum_g \sum_\ell e_{ig}e_{i\ell}p_ip_gq_\ell \tag{5.36}$$

$$F_{w1b} = E\{C_{w1} C_b\} = E\{(\sum_i c_ip_i)(\sum_i \sum_\ell e_{i\ell}p_iq_\ell)\} \tag{5.37}$$

$$F_{w2b} = E\{C_{w2} C_b\} = E\{(\sum_i \sum_j c_{ij} p_i p_j)(\sum_i \sum_\ell e_{i\ell} p_i q_\ell)\} \qquad (5.38)$$

$$F_{b22} = E\{C_b^2\} = E\{(\sum_i \sum_\ell e_{i\ell} p_i q_\ell)^2\} . \qquad (5.39)$$

From the above definition, the variance and covariance may be expressed as follows. They are decomposed into two parts; among-chromosome components and among-population components, i. e. the variance and covariance among the gene families within a population and those among population means under the same condition. This is because the population mean itself fluctuates from time to time due to finite population size.

Among chromosome components within a population:

$$\sigma_{w11}^2 = E_{w11} - F_{w11} \qquad \text{(variance of } c_i\text{)} \qquad (5.40)$$

$$\sigma_{w12}^2 = E_{w12} - F_{w12} \qquad \text{(covariance of } c_i \text{ and } c_{ij}\text{)} \qquad (5.41)$$

and

$$\sigma_{w22}^2 = E_{w22} - F_{w22} \qquad \text{(variance of } c_{ij}\text{)} \qquad (5.42)$$

Among chromosome components involving the identity coefficients between the populations:

$$\sigma_{w1b} = E_{w1b} - F_{w1b} \qquad \text{(covariance of } c_i \text{ and } e_{i\ell}\text{)} \qquad (5.43)$$

$$\sigma_{w2b} = E_{w2b} - F_{w2b} \qquad \text{(covariance of } c_{ij} \text{ and } e_{i\ell}\text{)} \qquad (5.44)$$

and

$$\sigma_{b22}^2 = E_{b22} - F_{b22} \qquad \text{(variance of } e_{i\ell}\text{)} . \qquad (5.45)$$

Among population components:

$$\sigma_{11}^2 = F_{w11} - \hat{C}_{w1}^2 \qquad \text{(variance of } C_{w1}\text{)} \qquad (5.46)$$

$$\sigma_{12} = F_{w12} - \hat{C}_{w1}\hat{C}_{w2} \qquad \text{(covariance of } C_{w1} \text{ and } C_{w2}\text{)} \qquad (5.47)$$

$$\sigma_{22}^2 \;=\; F_{w22} - \hat{C}_{w2}^2 \qquad \text{(variance of } C_{w2}) \qquad\qquad (5.48)$$

$$\sigma_{1b} \;=\; F_{w1b} - \hat{C}_{w1}\hat{C}_b \qquad \text{(covariance of } C_{w1} \text{ and } C_b) \qquad (5.49)$$

$$\sigma_{2b} \;=\; F_{w2b} - \hat{C}_{w2}\hat{C}_b \qquad \text{(covariance of } C_{w2} \text{ and } C_b) \qquad (5.50)$$

and

$$\sigma_{bb}^2 \;=\; F_{b22} - \hat{C}_b^2 \qquad \text{(variance of } C_b) \;. \qquad\qquad (5.51)$$

Note that when identity coefficients are available, the among population components may be applicable to the comparison between species, such as satellite DNA similarity between related species (Hennig and Walker 1970, Fry and Salser 1977). On the other hand, the among chromosome components may be useful for analysing the more detailed data on gene similarity between the gene families on different chromosomes within a population. The among chromosome components also have an important bearing if natural selection operates on gene variation contained in each family (see chapter 8).

In the following, the equations for calculating the above moments are derived. I shall present here an example of a calculation. The changes of other moments may be similarly obtained. Let us start with calculating the change of E_{w1b} by unequal crossing-overs. The left-hand side of the equation (5.8) is the rate of change of E_{w1b} by letting

$$f \;=\; (\sum_k x_{i,k}^2)(\sum_k x_{i,k}y_{\ell,k}) \;=\; \sum_k x_{i,k}^3 y_{\ell,k} + \sum_k \sum_{m\neq k} x_{i,k}^2 x_{i,m} y_{\ell,m} \;.$$

By using the formulae (5.10 - 5.16) and by taking one generation as a unit of t, the right-hand side becomes,

$$E[L(E_{w1b})] \;=\; E\left[\frac{1}{2}\left\{ \sum_k 6x_{i,k}y_{\ell,k} + \sum_k \sum_{m\neq k} 2x_{i,m}y_{\ell,m} \right\} V(\Delta x_{i,k}) \right.$$

$$+ \left\{ \sum_k \sum_{m\neq k} 2x_{i,k}y_{\ell,m} \right\} W(\Delta x_{i,k}\Delta x_{i,m})$$

$$\left. + \left\{ \sum_k 3x_{i,k}^2 y_{\ell,k} + \sum_k \sum_{m\neq k} 2x_{i,k}x_{i,m}y_{\ell,m} \right\} M(\Delta x_{i,k}) \right.$$

$$+ \left\{ \sum_k x_{i,k}^3 \right\} M(\Delta y_{\ell,k}) + \left\{ \sum_k \sum_{m \neq k} x_{i,k}^2 y_{\ell,m} \right\} M(\Delta x_{i,m})$$

$$\left. + \left\{ \sum_k \sum_{m \neq k} x_{i,k}^2 x_{i,m} \right\} M(\Delta y_{\ell,m}) \right]$$

$$= \alpha\{C_b + 2D_{b2} - 3E_{wlb}\} + \alpha'\{D_{b2} + D_{b3} + C_b$$

$$- 2E_{w2b} - F_{wlb}\} + \frac{n\alpha'}{2}\{2E_{w2b} + F_{wlb} - 3E_{wlb}\} \ .$$

This is equal to the amount of change of E_{wlb} due to intra- and inter-chromosomal unequal crossing-over in one generation.

Next, let us consider the change of E_{wlb} due to random genetic drift again by using the equation (5.8). Since $E_{wlb} = \sum_i \sum_\ell c_i e_{i\ell} p_i q_\ell$ and $M(\Delta p_i) = M(\Delta q_\ell) = W(\Delta p_i \Delta q_\ell) = 0$, the change of E_{wlb} by random drift is zero. Through mutation in one generation, E_{wlb} is reduced by the rate, 4v. Finally, the effect of equal inter-chromosomal crossing-over is, by using Fig. (3.2) and the formula (3.9), to change E_{wlb} by the amount $(\beta/3)(E_{w2b} - E_{wlb})$ under the assumption of a uniform distribution of r.

The changes of the other moments may be similarly calculated. As to F_{wll}, F_{wl2}, F_{w22}, F_{wlb}, F_{w2b} and F_{b22}, however, the changes due to unequal and equal crossing-overs are obtained more directly, because they are expectations of the products of the second moments. For example, $F_{wlb} = E\{C_{wl}C_b\}$ changes by the amount $\Delta C_{wl} \times \hat{C}_b = \alpha\{\hat{C}_b - F_{wlb}\} + \alpha'\{\hat{C}_b + (n_g - 1)F_{w2b} - n_g F_{wlb}\}$ due to unequal crossing-over in one generation, in which ΔC_{wl} is the expected change of C_{wl} by unequal crossing-overs. This is because the expected change of C_b by unequal crossing-overs is zero.

The resulting equations for linking the moments from one generation to the next are given first for the cases concerning the single population. As in my previous reports (Ohta 1978b, 1979), let M, S, K and R be the coefficients of the equations in matrix notation for deriving the moments corresponding to mutation, sampling, contraction-expansion process and gene exchange process due to crossing-overs, respectively. I also designate, by subscripts D and E, the coefficients of the equations of the third and fourth moments respectively. In the following, I shall assume that the sequence of events in one generation is; unequal intra- and inter-chromosomal crossing-overs (contraction-expansion process) → equal inter-chromosomal crossing-overs (gene exchange process) → random drift → mutation. Let

$$D = \begin{bmatrix} D_{w1} \\ D_{w2} \\ D_{w3} \end{bmatrix}$$. This vector changes from D_t to D_{t+1} in one generation

according to the formula

$$D_{t+1} = M_D S_D R_D (K_D D_t + A_D) \tag{5.52}$$

in which matrices are, by letting I be the identity matrix,

$$M_D = (1 - v)^3 \tag{5.53}$$

$$S_D = I - \frac{1}{2N_e} \begin{bmatrix} 0 & 0 & 0 \\ -1 & 1 & 0 \\ 0 & -3 & 3 \end{bmatrix} \tag{5.54}$$

$$R_D = I - \beta' \begin{bmatrix} \frac{1}{2} & -\frac{1}{2} & 0 \\ 0 & \frac{1}{3} & -\frac{1}{3} \\ 0 & 0 & 0 \end{bmatrix} \tag{5.55}$$

$$K_D = I - \alpha \begin{bmatrix} 3 & 0 & 0 \\ 0 & 1 & 0 \\ 0 & 0 & 0 \end{bmatrix} - \alpha' \begin{bmatrix} \frac{3n_g}{2} & -\frac{3n_g}{2} + 3 & 0 \\ 0 & n_g & -(n_g-1) \\ 0 & 0 & 0 \end{bmatrix} \tag{5.56}$$

and

$$A_D = \alpha \begin{bmatrix} 3C_{w1} \\ C_{w2} \\ 0 \end{bmatrix} + \alpha' \begin{bmatrix} \frac{3}{2}(C_{w1} + C_{w2}) \\ C_{w2} \\ 0 \end{bmatrix} . \tag{5.57}$$

where $\beta' = \beta + \gamma'\{1 - 3\mu'\}$, lower moments in A_D are those at the previous generation (t), and all parameters v, $1/2N_e$, β', α and $n_g\alpha'$ are assumed to be much less than unity. Note that β' represents the total effect of gene exchanges between the families due to equal and unequal inter-chromosomal crossing-overs, since β is the equal cross-ing-over rate and γ' is the unequal crossing-over rate with mean shift of m' genes. The term 3μ' is a correction factor as in equation (3.20), page 32.

The vector of the fourth moments,

$$
E = \begin{bmatrix}
E_{w11} \\
E_{w12} \\
E_{w22} \\
E_{w3} \\
F_{w11} \\
F_{w12} \\
F_{w22} \\
F_{w2}
\end{bmatrix}
$$

may be similarly generated. The following equation gives the change of the vector in one generation, again by assuming all parameters v, $1/2N_e$, β', α and $n_g\alpha'$ are much less than unity.

$$
E_{t+1} = M_E S_E R_E (K_E E_t + A_E) \tag{5.58}
$$

in which

$$
M_E = (1 - v)^4 \tag{5.59}
$$

$$
S_E = I - \frac{1}{2N_e}
\begin{bmatrix}
0 & 0 & 0 & 0 & & & & \\
-1 & 1 & 0 & 0 & & & & \\
-1 & 0 & 1 & 0 & & \mathbf{O} & & \\
0 & -2 & -1 & 3 & & & & \\
-1 & 0 & 0 & 0 & 1 & 0 & 0 & 0 \\
0 & -2 & 0 & 0 & -1 & 3 & 0 & 0 \\
0 & 0 & 0 & 0 & 0 & -2 & 6 & -4 \\
0 & -2 & -1 & 0 & 0 & 0 & 0 & 3
\end{bmatrix}
\tag{5.60}
$$

$$
R_E \;=\; I - \beta'
\begin{bmatrix}
\frac{8}{15} & -\frac{6}{15} & -\frac{2}{15} & 0 & & & & \\
0 & 0 & 0 & 0 & & & & \\
0 & 0 & 0 & 0 & & & \bigcirc & \\
0 & 0 & 0 & 0 & & & & \\
& & & & \frac{2}{3} & -\frac{2}{3} & 0 & 0 \\
& \bigcirc & & & 0 & \frac{1}{3} & -\frac{1}{3} & 0 \\
& & & & 0 & 0 & 0 & 0 \\
& & & & 0 & 0 & 0 & 0
\end{bmatrix}
\tag{5.61}
$$

with $\beta' \;=\; \beta + \gamma'(1 - 3\mu')$.

$$
K_E \;=\; I - \alpha
\begin{bmatrix}
6 & 0 & 0 & 0 & & & & \\
0 & 3 & 0 & 0 & & & & \\
0 & 0 & 2 & 0 & & & \bigcirc & \\
0 & 0 & 0 & 1 & & & & \\
& & & & 2 & 0 & 0 & 0 \\
& \bigcirc & & & 0 & 1 & 0 & 0 \\
& & & & 0 & 0 & 0 & 0 \\
& & & & 0 & 0 & 0 & 0
\end{bmatrix}
\tag{5.62}
$$

$$
- \,\alpha'
\begin{bmatrix}
2n_g & 6 & 0 & 0 & 0 & 0 & 0 & 0 \\
0 & \frac{3}{2}n_g & 0 & -(n_g-2) & 0 & -\frac{n_g-2}{2} & 0 & 0 \\
0 & 0 & 2n_g & -2(n_g-1) & 0 & 0 & 0 & 0 \\
0 & 0 & 0 & n_g & 0 & 0 & -(n_g-1) & 0 \\
& & & & 2n_g & -2(n_g-1) & 0 & 0 \\
& & \bigcirc & & 0 & n_g & -(n_g-1) & 0 \\
& & & & 0 & 0 & 0 & 0 \\
& & & & 0 & 0 & 0 & 0
\end{bmatrix}
$$

and

$$
A_E = \alpha \begin{bmatrix} 2C_{w1} + 4D_{w1} \\ C_{w1} + 2D_{w1} \\ 2D_{w1} \\ D_{w3} \\ 2C_{w1} \\ C_{w2} \\ 0 \\ 0 \end{bmatrix} + \alpha' \begin{bmatrix} 2(n_g+1)D_{w2} + 2(C_{w1}+D_{w1}) \\ C_{w2} + D_{w2} + D_{w3} \\ 2D_{w1} \\ D_{w2} \\ 2C_{w1} \\ C_{w2} \\ 0 \\ 0 \end{bmatrix} \tag{5.63}
$$

with $\alpha = m\gamma/n_g^2$ and $\alpha' = m'\gamma'/n_g^2$ and various lower moments in A_E are those at the previous generation (t).

By using the equations (5.52 - 5.63), it is possible to generate the variance and covariance of identity coefficients from one generation to the next. An interesting question would be the extent of identity coefficients fluctuations when equilibrium is attained among various forces. The variance and covariance at equilibrium were numerically obtained for some interesting cases (Table 5.1). It can be seen from the table that the identity coefficients are generally accompanied by large variances and therefore they fluctuate greatly over time.

When there is no inter-chromosomal crossing-over ($\beta = 0$, $\gamma' = 0$), the formulas become simpler. It is possible to express the variances of c_i and C_{w1} by the following equation (Ohta 1978b).

$$
\sigma_{w11}^2 = E_{w11} - F_{w11} = \frac{16N_e v}{(1+\theta)(2+\theta)(3+\theta)(1+4N_e\alpha+8N_e v)} \tag{5.64}
$$

and

$$
\sigma_{11}^2 = F_{w11} - \hat{C}_{w1}^2 = \frac{2\theta}{(1+\theta)^2(2+\theta)(3+\theta)} - \sigma_{w11}^2 \tag{5.65}
$$

where $\theta = 2v/\alpha$. When $N_e \to \infty$, the upper formula (5.64) reduces to the first part of the right hand side of the equation (5.65), $2\theta/\{(1+\theta)^2(2+\theta)(3+\theta)\}$, which is the same as the formula for the variance of homozygosity at a locus when mutation and random drift balance each other (Stewart 1976). In our case, the balance is between

63

Table 5.1 The variance and covariance of identity coefficients at equilibrium.

Parameters			N_e	C_{w1}	C_{w2}	Between-population components			Within-population components		
α	β'	α'				σ_{11}^2	σ_{12}	σ_{22}^2	σ_{w11}^2	σ_{w12}	σ_{w22}^2
10^{-3}	0	0	500	0.91	0.83	0.008	0.013	0.029	0.018	0.010	0.047
10^{-3}	0	0	2500	0.91	0.61	0.002	0.004	0.032	0.023	0.013	0.152
10^{-3}	10^{-2}	0	500	0.71	0.65	0.049	0.062	0.074	0.021	0.022	0.022
10^{-2}	10^{-1}	10^{-3}	100	0.91	0.89	0.027	0.033	0.048	0.008	0.008	0.007
10^{-3}	10^{-1}	10^{-3}	100	0.64	0.62	0.075	0.078	0.080	0.007	0.006	0.002
0	10^{-1}	10^{-3}	100	0.47	0.46	0.068	0.069	0.070	0.005	0.004	0.001

(Other parameters; ν = 5 × 10^{-5}, μ' = 0.125 and n_g = 20)

mutation and unequal (intra-chromosomal) crossing-over and the between-population component of variance (σ_{11}^2) disappears, when $N_e \to \infty$.

Next, the equations for generating the variance and the covariance of the identity coefficients involving the between-population
identity (formulas 5.43 - 5.45 and 5.49 - 5.51) are presented. The
identity coefficient between the two isolated populations, C_b, decreases on the average with the constant rate, $2v$, each generation
(equation 5.7). However individual cases may take quite different
paths and one can infer, from the variance of C_b, how much they may
deviate from the prediction. Let us denote the vector of the third
moments by

$$d = \begin{bmatrix} D_{w1b} \\ D_{w2b} \end{bmatrix}$$

This vector changes from d_t to d_{t+1} in one generation, according to
the following equation, and by again letting M, S, K and R be the
coefficients of the matrix equations for deriving the moments corresponding to mutation, sampling, unequal crossing-over (contraction-
expansion process) and equal crossing-over (gene exchange process).

$$d_{t+1} = M_d S_d R_d (K_d d_t + A_d) \tag{5.66}$$

in which the matrices are

$$M_d = (1 - v)^3 \tag{5.67}$$

$$S_d = I - \frac{1}{2N_e} \begin{bmatrix} 0 & 0 \\ -1 & 1 \end{bmatrix} \tag{5.68}$$

$$R_d = I - \beta' \begin{bmatrix} \frac{1}{3} & -\frac{1}{3} \\ 0 & 0 \end{bmatrix} \tag{5.69}$$

$$K_d = I - \alpha \begin{bmatrix} -1 & 0 \\ 0 & 0 \end{bmatrix} - \alpha' \begin{bmatrix} n_g & -(n_g-1) \\ 0 & 0 \end{bmatrix} \qquad (5.70)$$

and

$$A_d = (\alpha + \alpha') \begin{bmatrix} C_b(t) \\ 0 \end{bmatrix} \qquad (5.71)$$

where $\beta' = \beta + \gamma'\{1 - 3\mu'\}$ as before. Also all parameters v, $1/2N_e$, α, $n_g\alpha'$ and β are again assumed to be much less than unity.

Next let $e = \begin{bmatrix} E_{w1b} \\ E_{w2b} \\ E_{b2} \\ E_{b3} \\ F_{w1b} \\ F_{w2b} \\ F_{b2} \end{bmatrix}$. The vector e changes in one genera-

tion following the equation,

$$e_{t+1} = M_e S_e R_e (K_e e_t + A_e) \qquad (5.72)$$

where

$$A_e = \alpha \begin{bmatrix} C_b + 2D_{w1b} \\ D_{w2b} \\ 2D_{w1b} \\ D_{w2b} \\ C_b \\ 0 \\ 0 \end{bmatrix} + \alpha' \begin{bmatrix} C_b + D_{w1b} + D_{w2b} \\ (D_{w1b} + D_{w2b})/2 \\ 2D_{w1b} \\ D_{w2b} \\ C_b \\ 0 \\ 0 \end{bmatrix} \qquad (5.73)$$

$$M_e = (1 - v)^4 \qquad (5.74)$$

$$S_e = I - \frac{1}{2N_e} \begin{bmatrix} 0 & 0 & 0 & 0 & & & \\ -1 & 1 & 0 & 0 & & & \\ 0 & 0 & 0 & 0 & & \bigcirc & \\ 0 & 0 & -1 & 1 & & & \\ -1 & 0 & 0 & 0 & 1 & 0 & 0 \\ 0 & -2 & 0 & 0 & -1 & 3 & 0 \\ 0 & 0 & 0 & -2 & 0 & 0 & 2 \end{bmatrix} \qquad (5.75)$$

$$K_e = I - \alpha \begin{bmatrix} 3 & 0 & 0 & 0 & & & \\ 0 & 1 & 0 & 0 & & \bigcirc & \\ 0 & 0 & 2 & 0 & & & \\ 0 & 0 & 0 & 1 & & & \\ & & & & 1 & 0 & 0 \\ & \bigcirc & & & 0 & 0 & 0 \\ & & & & 0 & 0 & 0 \end{bmatrix}$$

$$- \alpha' \begin{bmatrix} \frac{3}{2}n_g & -(n_g-2) & 0 & 0 & -(\frac{n_g}{2}-1) & 0 & 0 \\ 0 & (n_g+\frac{1}{2}) & 0 & 0 & \frac{1}{2} & -n_g & 0 \\ 0 & 0 & 2n_g & -2(n_g-1) & 0 & 0 & 0 \\ 0 & 0 & 0 & n_g & 0 & 0 & -(n_g-1) \\ & & & & n_g & -(n_g-1) & 0 \\ & & \bigcirc & & 0 & 0 & 0 \\ & & & & 0 & 0 & 0 \end{bmatrix} \qquad (5.76)$$

and

$$R_e = I - \beta' \begin{bmatrix} \frac{1}{3} & -\frac{1}{3} & 0 & 0 & & & & \\ 0 & 0 & 0 & 0 & & O & & \\ 0 & 0 & 0 & 0 & & & & \\ 0 & 0 & 0 & 0 & & & & \\ & & & & & \frac{1}{3} & -\frac{1}{3} & 0 \\ & O & & & & 0 & 0 & 0 \\ & & & & & 0 & 0 & 0 \end{bmatrix} \tag{5.77}$$

The lower moments in (5.73) are those at the t-th generation.

5.3 Numerical examples and some applications

It may be interesting to clarify the relationship between the average identity probability (C_b) and variance (σ_{bb}^2; variance of C_b) or covariance (σ_{w2b}; covariance of C_b and C_{w2}) under various conditions. $\hat{C}_b$ reduces every generation at the rate 2v (formula 5.7). However it merely represents the average identity probability over many species comparisons and individual cases may deviate tremendously from the mean. Based on the magnitude of σ_{bb}^2, it is possible to predict how much they may deviate from the mean. The prediction may be applied to the pattern of genetic differentiation of satellite DNA in some mammalian species (Fry and Salser 1977). This example is particularly interesting, since, from the highly preserved satellite DNA of some species comparisons, the unique importance of this satellite for the species is seriously considered (Fry and Salser 1977, Dover 1978).

By using the theory developed here various components of the variance and covariance are numerically obtained. In general, when $N_e v$ is small, the gene families become uniform in one population and among-chromosome components become small as compared with among-population components. Such relationships may be seen from the formulas (5.64, 5.65) although the formulas are restricted to the case of no inter-chromosomal crossing-over.

We are considering a simple situation in which a population is split into two identical ones with the same effective size (N_e) as the original one and the original population is in equilibrium at the time of splitting. Therefore all moments of identity probabilities between the two populations are equal to the corresponding moments of a single

equilibrium population at the time of splitting. The equilibrium
moments of a single population were calculated by using the equations
(5.52 - 5.63). Starting from the equilibrium moments, the variance
and the covariance of the identity coefficient involving the between-
population identity is generated from one generation to the next by
using the equations (5.66 - 5.77). Here I assume the sequence of
events in one generation is unequal crossing-over (contraction-expansion process) → equal crossing-over (gene exchange process) → random
drift → mutation.

Table 5.2 gives some numerical examples. Both among-chromosome
and among-population variance components for some values of $\hat{C}_b$ are
given to show their relative magnitude. Note that $\hat{C}_b$ is equal to $\hat{C}_{w2}$
at the time of splitting of the two species and decreases by the constant rate, 2v. Therefore the time since divergence in each case
differs for the same value of $\hat{C}_b$.

The among-population components of the variance and covariance
were studied in more detail. Results are presented in Table 5.3. An
interesting general picture emerges from these analyses; σ_{bb}^2 becomes
large when the average identity probability within the species is high,
whereas it does not become so large when the gene family is originally
heterogeneous within the population. In other words, the differentiation of the family with uniform gene members is accompanied by a large
statistical variance. The situation is analogous to the differentiation of selectively neutral enzyme loci (Li and Nei 1975). Thus, for
gene families with uniform members, the differentiation may be quite
rapid in some cases and slow in others purely by chance even if the
parameters are the same.

Since the parameters are not known in the case of satellite DNA
of rodents, detailed calculations can not be done at the moment for
this particular example. Another difficulty of the model is the
assumption that the mean gene (unit) number per family (n_g) stays
constant, whereas it actually varies during the evolution of satellite
DNA. However, one may regard the process as follows; the parameters
α and α' are, roughly speaking, the rates of approach to homogeneity
of the gene members in the family due to unequal crossing-over, whereas v increases heterogeneity, and N_e and β control the heterogeneity
among different chromosomes in the population. Then the differentiation pattern of gene families between isolated populations would
roughly follow the prediction of our simple model as far as identity
probability is concerned, even when the mean family size does not stay
constant.

Table 5.2 The variance and covariance of identity coefficients involving that between two isolated populations.

	$\hat{C}_b$	0.7	0.6	0.5	0.4	0.3	0.2	0.1
$\hat{C}_{w1} = 0.82$	σ_{1b}	0.007	0.005	0.004	0.004	0.003	0.002	0.001
	σ_{2b}	0.004	0.003	0.003	0.002	0.002	0.001	0.001
$\hat{C}_{w2} = 0.79$	σ_{bb}^2	0.018	0.018	0.015	0.012	0.009	0.006	0.003
	σ_{w1b}	0.030	0.024	0.020	0.016	0.012	0.008	0.004
$N_e v = 10^{-2}$	σ_{w2b}	0.036	0.030	0.025	0.020	0.015	0.010	0.005
	σ_{b22}^2	0.066	0.120	0.149	0.156	0.145	0.116	0.067
$\hat{C}_{w1} = 0.80$	σ_{1b}		0.016	0.013	0.011	0.008	0.005	0.003
	σ_{2b}		0.015	0.012	0.010	0.007	0.005	0.002
$\hat{C}_{w2} = 0.67$	σ_{bb}^2		0.068	0.065	0.056	0.042	0.028	0.014
	σ_{w1b}		0.021	0.016	0.013	0.010	0.006	0.003
$N_e v = 5\times10^{-2}$	σ_{w2b}		0.042	0.033	0.027	0.020	0.013	0.007
	σ_{b22}^2		0.059	0.089	0.106	0.109	0.093	0.056
$\hat{C}_{w1} = 0.73$	σ_{1b}					0.014	0.009	0.004
	σ_{2b}					0.022	0.014	0.007
$\hat{C}_{w2} = 0.37$	σ_{bb}^2					0.109	0.081	0.042
	σ_{w1b}					0.007	0.004	0.002
$N_e v = 0.25$	σ_{w2b}					0.022	0.015	0.007
	σ_{b22}^2					0.028	0.031	0.024

(Parameters; $\alpha = 10^{-3}$, $\alpha' = 0$, $\beta' = 10^{-3}$ and $v = 10^{-4}$)

Table 5.3 The variance of identity coefficient between the two isolated populations (σ_{bb}^2).

α	α'	v	β'	$\hat{C}_{w2}/\hat{C}_b$	0.8	0.7	0.6	0.5	0.4	0.3	0.2	0.1
10^{-3}	0	10^{-4}	10^{-3}	0.79		0.072	0.116	0.145	0.156	0.146	0.118	0.069
10^{-3}	0	5×10^{-4}	10^{-3}	0.41					0.030	0.034	0.035	0.026
0	10^{-3}	10^{-5}	10^{-1}	0.81	0.065	0.106	0.151	0.176	0.181	0.165	0.130	0.075
0	10^{-3}	5×10^{-5}	10^{-1}	0.46					0.063	0.065	0.060	0.041
10^{-3}	10^{-3}	10^{-5}	10^{-1}	0.89	0.096	0.153	0.192	0.210	0.208	0.186	0.143	0.083
10^{-3}	10^{-3}	5×10^{-5}	10^{-1}	0.62			0.079	0.093	0.113	0.113	0.095	0.058
10^{-3}	10^{-3}	10^{-4}	10^{-1}	0.45					0.061	0.063	0.059	0.040

(Other parameters; $N_e = 10^2$, $n_g = 20$, and $\mu' = 0.125$)

It has been shown that the four species (kangaroo rat, guinea pig, pocket gopher and antelope ground squirrel) have the same α satellite DNA (Fry and Salser 1977), whereas some of the more closely related species to one of these four have apparently lost the sequence (Hennig and Walker 1970). Thus the differentiation of α satellite is very rapid in some species but quite slow in others. Fry and Salser argue from this fact that the sequence may have some specific adaptive meaning for speciation. It is further argued that if the rate of divergence of satellite DNA is 10^{-8} per base pair per year for rodents (Southern 1974), the sequences of satellites of these four species should have changed 40 - 50% of their bases (Fry and Salser 1977). This prediction would apply only when many species comparisons are made and the average is examined. The mean identity probability of homologous bases of the two species would be $e^{-0.9} \approx 0.4$ (by letting vt = 0.45 in the formula 5.7) of the corresponding identity probability within the species. However, the differentiation of this satellite would be accompanied by a large statistical deviation and Fry and Salser's observation may be a very likely outcome of a stochastic differentiation.

Application of the theory to other multigene families such as ribosomal RNA or transfer RNA gene families awaits future reports such as those by Wellauer et al. (1976) and Krystal and Arnheim (1978). Here our model fits better than in the case of satellite DNA because of the fairly stable family size. Amino acid sequence data of variable regions of immunoglobulins provide still more interesting examples of the evolution of multigene families for the application of the theory. Statistical analyses on sequence variability of immunoglobulins will be reported in the next chapter.

CHAPTER 6

STATISTICAL ANALYSES ON SEQUENCE VARIABILITY

OF IMMUNOGLOBULINS

As discussed in chapter 1, the antibody gene family represents a most interesting example of evolving multigene families. It is particularly interesting because of its enormous diversity within and between species. It will be shown in this chapter that amino acid diversity of variable regions of immunoglobulins fits very well to the prediction of the simple model discussed in the previous chapters.

6.1 Average identity coefficient

Let us measure the similarity of variable region sequences of immunoglobulins by the identity coefficient studied in the previous chapters. Instead of taking the whole variable region as a unit for the comparison of identity, let us fix our attention on a particular residue position inside each gene and compare the amino acid identities between corresponding positions in gene members within the family. If I take the average of identity coefficients at each position over a sufficient number of sites, the average values may be compared with the theoretical predictions, $\hat{C}_{w1}$, $\hat{C}_{w2}$ and $\hat{C}_{b}$ (equations 3.22, 3.23 and 5.7). I have reported the preliminary analyses on amino acid identity, by separately treating the hypervariable and the framework regions (for the two regions, see Chapter 1) and have shown that hypervariability may readily be explained as genetic variation by assuming that the mutation rate (more precisely, the rate of amino acid substitution in evolution) is roughly three times higher at the hypervariable region than at the framework region (Ohta 1978c). In other words, the hypervariability is a necessary consequence of a high mutation rate as predicted by the formulae (3.22) and (3.23).

The identity coefficient at each residue position is calculated from the compilation by Kabat et al. (1976) of reported sequences of variable regions of immunoglobulins. The mean identity coefficient over amino acid sites of a particular type (such as κ, λ or heavy) of immunoglobulin in one species corresponds to C_{w2} of the theoretical model. Note here that, as far as the amino acid site is concerned,

$C_{w1} \approx C_{w2}$ from the equation (3.23), since $4N_e v$ is generally much less than unity. Throughout this chapter, I use C_w instead of C_{w1} or C_{w2}. The identity coefficient of amino acids between two species (such as human and mouse) corresponding to C_b is also estimated. These identity coefficients may be calculated by the following method. The frequencies of 20 kinds of amino acid are counted for each residue position of the reported sequences of a given type of immunoglobulin of a species (such as human κ sequences). The sum of squares of these frequencies gives the estimated value of C_w at each position of that type. The formula may be written as follows.

$$C_{w,r} = \sum_i g_{i,r}^2 \qquad (6.1)$$

in which the subscript r denotes the residue position, $g_{i,r}$ is the frequency of the i-th amino acid (i = 1 ∿ 20) at that position and summation is over all i. Similarly, $C_{b,r}$ may be estimated by calculating the sum of cross products of amino acid frequencies at homologous residue positions of a type of immunoglobulin of the two species. The formula may be written as follows

$$C_{b,r} = \sum_i g_{i,r} h_{i,r} \qquad (6.2)$$

where $g_{i,r}$ and $h_{i,r}$ are the frequencies of the i-th amino acid at the r-th position of a given type of immunoglobulin of the first and the second species respectively. The mean identity coefficient may be obtained by calculating the average of the above values over a suitable set of r;

$$\bar{C}_w = \frac{\sum_r C_{w,r}}{n_s} \qquad (6.3)$$

and

$$\bar{C}_b = \frac{\sum_r C_{b,r}}{n_s} \qquad (6.4)$$

where n_s is the number of amino acid sites at which data are available in a given set of r. The above procedure is analogous to that of

estimating homozygosity and gene identity of related species at enzyme
loci in population genetics (see Nei 1975).

The average identity coefficients are obtained separately for the
hypervariable and the framework regions. According to the numbering
scheme of Kabat et al. (1976), the hypervariable regions include
positions 24-34, 50-56 and 89-97 in the light chain and 31-35, 50-65,
81-85 and 95-102 in the heavy chain. The remaining sites are the
framework regions. Among the sequences compiled by Kabat et al.
(1976), immunoglobulin chains for which only a small part is sequenced
are not used. Positions with deletions or ambiguous amino acids are
omitted from the calculation. Thus, the sequences used are; the heavy
chain: EU, DI, DAW, OU, COR, HE, NEWM, NIE, TEI, WAS, JON, LAY, POM,
GAL ZAP, TUR, TRO and GA of human, MOPC 315, MOPC 21, MOPC 173, HOPC
8, TEPC 15, S 107, MCPC 603, MOPC 167, MOPC 511 and W 3207 of mouse
and BS-5, K-25, 2690, AA1POOL and BS-1 of rabbit, the κ chain: AG,
DEE, HAU, OU, ROY, BEL, EU, CAR, AU, SCW, BI, BJ 26, BJ 19, GAL (1),
REI, NI, KA, TEW, MIL, CUM, TI, FR 4, RAD and LEN of human, MOPC 41,
MOPC 173, MOPC 70, MOPC 321, TEPC 124, MOPC 63, MOPC 21, MOPC 460,
MOPC 511 and MPC 11 of mouse and 4135, BS-1, BS-5, K-25, 2717, 3315,
3368, 3374, 120, 3547, K4820, K30-267, K16-167, K27-489, K29-213,
K9-335 and K9-338 of rabbit, the λ chain: HA, NEW, VOR, NEWM, VIL,
NEI, BOH, TRO, X, BAU, KERN, DEL, SH, BO and MCG of human and MOPC
104E, RPC 20, J558, XS 104, SAPC 178, SAPC 176, Y5606, Y5444, HOPC
2020 and MOPC 315 of mouse.

Figure 6.1 shows the average identity coefficient of the heavy
chain and the κ chain. The within-species identity coefficients ($\bar{C}_w$)
are given in the solid circle and the between-species values ($\bar{C}_b$) are
given in the broken circle. In both circles, the upper number is the
identity coefficient at the hypervariable region, and the lower number,
that at the framework region. From the figure, it is clear that the
identity coefficient of the framework region and that of the hypervariable region vary correlatively. Following my report (Ohta 1980a),
let us examine the relationship between the two identity coefficients
in more detail. Alternative major hypotheses are either that the
hypervariability arises from a higher rate of mutations being fixed in
that part of the gene coding for the hypervariable regions compared to
the framework coding regions of the gene (the germ line hypothesis) or
that the dominant source of extra variability in the hypervariable
regions are somatic mutations superimposed upon a germ line fixation
rate that is approximately uniform over the entire gene (the somatic
mutation hypothesis). Later, I shall also examine some intermediate

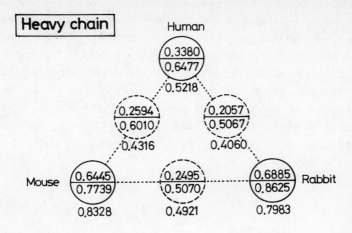

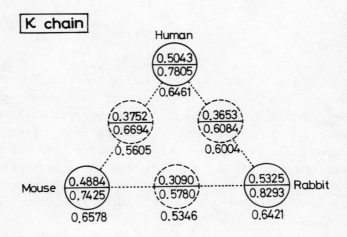

Figure 6.1 Average identity coefficient of amino
acid per residue position. Closed
circles, identity coefficient within
a species (C_w); broken circles, that
between species (C_b); upper number,
hypervariable regions; and the lower
number, framework regions. The numbers
below the circle show the ratio of
identity coefficients of hypervariable
regions to those of the framework
regions.

hypotheses between the above two extremes. Let us denote the identity coefficient at the framework region and that at the hypervariable region by C^{FW} and C^{HV} respectively. Then under the somatic mutation hypothesis, the lower value of C^{HV} compared to C^{FW} depends upon the added contribution of the somatic changes which depend neither upon time since divergence of the gene families, nor upon the gene family size. On the other hand, in the germ line hypothesis, divergence increases at the basic differential rate in both hypervariable and in framework regions.

From equation (5.7), we have

$$C_b^{FW}(t) = C_b^{FW}(0)\exp\{-2v^{FW}t\} \qquad (6.5)$$

and

$$C_b^{HV}(t) = C_b^{HV}(0)\exp\{-2v^{HV}t\} \qquad (6.6)$$

in which $\exp\{\cdot\}$ represents the exponential function and v^{FW} and v^{HV} are respectively the mutation rate (evolutionary rate) <u>per amino acid site per year</u> in the framework and in the hypervariable regions when t is measured in years. Note that, without natural selection, the mutation rate becomes equal to the evolutionary rate for conventional genetic loci (Kimura 1968). The situation is more complicated for multigene families, yet the two rates are equal if no systematic pressure is involved. The effect of natural selection will be discussed in Chapter 8.

Actually, the within-population identity coefficient C_w varies considerably from species to species. The main cause would be the change in family size (n_g) during evolution. For example, the gene family size of the mouse lambda chain should have greatly contracted since its divergence from the other families, and the diversity of this family is very limited. It can be seen from the equation (3.22), that the within-population identity coefficient C_w is lowered by the mutation rate v with its coefficient positively correlated with the family size n_g. Thus, one would expect, even if not identical, a similar relationship between different values of C_w as in equations (6.5) and (6.6).

Figure 6.2 shows the regression line of $Y = -\ln C^{HV}$ on $X = -\ln C^{FW}$ which is $Y = 2.586X + 0.009$. Each point represents an observed value which may be identified by a letter or a number as

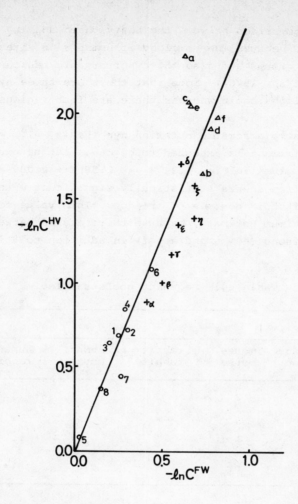

Figure 6.2 Relationship between $-\ln C^{FW}$ and $-\ln C^{HV}$. The straight line represents the regression line, and each dot represents an observed value which may be identified by the symbols listed in Table 6.1.

summarized in Table 6.1. The open circles (1 ∿ 8) are the within-population values, the crosses (α ∿ η) are the between-species values and the triangles (a ∿ f) are the comparisons between the κ- and the

λ-chain. The comparison between the heavy chain and the light chain
are not included because the homology of amino acid sites is ambiguous
between the two groups and also the hypervariable regions are different (Kabat et al. 1976). Note that there are three hypervariable
regions in the light chain whereas there are four regions in the heavy
chain.

Under the simple somatic mutation hypothesis, $v^{FW} = v^{HV}$, the
regression coefficient is expected to be one. Let us examine this
hypothesis. As given in Fig. 6.2, Y = 2.586X + 0.009 and the deviation of 2.586 from one is statistically significant with probability
of about 0.01 with six degrees of freedom. This value for degrees of
freedom is a minimum estimate, because there are 13 between-protein
families comparisons ($\alpha \sim \eta$ and $a \sim f$) in addition to 8 independent

Table 6.1 Sequence pools sampled.

Protein family	Taxa					
	human human	mouse mouse	rabbit rabbit	human mouse	human rabbit	mouse rabbit
$\kappa-\kappa$	1	2	3	α	β	γ
$\lambda-\lambda$	4	5	*	δ	*	*
h-h	6	7	8	ϵ	ζ	η
$\kappa-\lambda$	a	b	*	c,d	e,*	f,*

Each sampling requires two drawings which are identifiable
according to the taxa and protein family as indicated by desig-
nation in the body of the table. These designations identify
the points in the figure. For the protein families, κ, λ and h
denote kappa, lambda and heavy chains of immunoglobulins respectively. An asterisk, *, shows a missing comparison that
occurs because the rabbit has no lambda chains. The group given
by numbers are the "within" comparisons involving a single population (one protein family, one taxon) and are shown as circles
(o) in the figure. The group given by Greek letters are
comparisons in a single family of proteins but between different
taxa and are shown as plusses (+) in the figure. The group in
the bottom line of the table involves comparisons between κ
and λ families of protein and are shown as triangles (Δ) in the
figure.

within-population values. Therefore, the above probability of 0.01 is
a maximum estimate. At any rate, the higher basic rate of mutant

accumulation at the hypervariable region as compared to that at the framework region is shown to be statistically significant and the straight line gives a very good fit to the observed values. In other words, the differentiation pattern of immunoglobulins follows the prediction of the germ line theory, and is incompatible with the simple somatic mutation theory. Note, however, that one residue position at the edge of the J segment (96th site of the light chain by Kabat et al.'s numbering scheme, see Figs. 1.4 and 1.5) may be an exception at which extra-variability of amino acids may be generated somatically through the joining process of V gene and J segment, since the recombination site may be modulated (Sakano et al. 1979, Seidman et al. 1979).

It is interesting to examine how the identity coefficient between the species relates to the divergence time in some detail. I have estimated, based on equations (6.5) and (6.6), 2vt, a quantity equivalent to genetic distance at enzyme loci (Nei 1972) for various comparisons of κ-chains or heavy-chains (Table 1 of Ohta 1978c). The quantity 2vt may be estimated by substracting $-\ell n\ C_w$ from $-\ell n C_b$. In my previous calculation, individual values were used for $-\ell n\ C_w$ or $-\ell n\ C_b$. Here it is more reasonable to use average values. This is because individual identity coefficients fluctuate considerably from species to species, and species comparisons are more remote than the case of estimating genetic distance at enzyme loci and hence it is unlikely that various parameters such as population size or family size remain unchanged since divergence. The calculation was done only for the framework regions and Table 6.2 gives the summary of the calculation. From the Table 6.2, v turns out to be 2.03×10^{-9} per year if we assume that the lagomorphs split 90 million years ago. Thus, the framework region is evolving by about the average rate of various proteins (Dayhoff 1972). As estimated by the regression analysis, v is 2.6 times larger at the hypervariable region than at the framework region, therefore v at the hypervariable region is roughly equal to the rate of fibrinopeptides which is the highest among the known rates (Dayhoff 1972). The difference of the evolutionary rates between the two regions is considered to reflect the structural constraints of the immunoglobulin molecule. In general, the stronger the constraint is, the lower is the evolutionary rate (Kimura and Ohta 1974). Novotný et al. (1977) fully discussed the relationship between the sequence variability and domain tertiary structure of λ chains. Also, from the result of Table 6.2, the divergence time of human and mouse may be estimated to be 66 million years, which is a reasonable estimate paleontologically, although a possible slowdown of molecular evolution in primates (e. g. Fitch 1973) makes the interpretation a little uncertain.

Table 6.2 Calculations on evolutionary rate per
amino acid site at the framework region.

Species Comparison	2vt	v per year
Human-mouse	0.2704	
Human-rabbit	0.3535	2.03×10^{-9}
Mouse-rabbit	0.3788	

6.2 Identity coefficient in association with subfamilies of variable region sequences

It is often found that one protein family such as human κ or mouse heavy chains consists of several distinct subfamilies, and between the members of subfamilies, recombination appears to be rather limited (Dayhoff 1972, Kabat et al. 1976). Thus variable region gene families apparently have subdivided structure on the chromosome. Then gene diversity within each subgroup needs to be investigated. In this section, I shall present quantitative analyses on identity coefficients of amino acids in each subgroup in relation to the theoretical predictions based on competing hypotheses.

The basic idea of the analysis is that under the germ line theory, the amino acid diversity of a subgroup depends mainly upon the number of genes (per genome) of the sequences belonging to the subgroup, whereas the added contribution of diversity due to somatic mutation is independent of the gene number and should be constant.

Let us denote the number of genes per genome of a subfamily by n_{sf}. Then one may use equation (3.22) to predict the identity coefficient within a subfamily, by assuming that equilibrium has been reached with respect to the identity coefficient of the subfamily. Actually, the assumption may not be satisfied. Then n_{sf} may be different from the actual gene number in the subfamily and it represents a value at some past time. For example, subfamily size may have recently expanded and the identity coefficient reflects the subfamily size of the past. In terms of using the equation (3.22), however, it does not matter whether or not n_{sf} represents the today's subfamily size. The situation is somewhat analogous to the use of an "effective population size" for analysing heterozygosity in population genetics when actual species either contract or expand through the evolutionary process (e. g. Kimura and Ohta 1971a).

The equation (3.22) takes the following form, by replacing n_g with n_{sf} and by noting $C_{w1} \approx C_{w2} = C_w$ under the assumption that four times the product of population size and mutation rate is much less than unity ($4N_e v_{aa} \ll 1$),

$$C_w = \frac{\alpha + \alpha'}{\alpha + \alpha' + 2v_{aa} + \dfrac{4N_e v_{aa}}{1 + 4N_e v_{aa}}\left\{\dfrac{\beta'}{3} + (n_{sf} - 1)\alpha'\right\}} \qquad (6.7)$$

in which v_{aa} now represents the mutation rate <u>per amino acid site per generation</u> and β' is the inter-chromosomal crossing-over rate (equal + corrected unequal) <u>per subfamily</u> per generation. As reasonable assumptions, let both the crossing-over rates (β, γ and γ', see page 22) and the mean gene number of shifts at unequal crossing-over (m and m', see page 22) be proportional to the subfamily size n_{sf}. Then α and α' become independent of n_{sf}, since $\alpha = m\gamma/n_{sf}^2$ and $\alpha' = m'\gamma'/n_{sf}^2$. By further noting $\beta' = \beta + \gamma'(1 - 3m'/n_{sf})$, the following relationship is obtained between C_w and n_{sf}.

$$C_w = \frac{1}{1 + (A + Bn_{sf})v_{aa}} \qquad (6.8)$$

in which A and B are constant.

It would be interesting to apply the above formula to the observed identity coefficients of subfamilies of variable region sequences. Let us examine the relationship among observed identity coefficients in view of the controversy on relative importance of germ line encoded variability and somatic mutation on hypervariability. The simplest somatic mutation hypothesis assumes that all extra-variability at the hypervariable region is due to somatic mutation (e. g. Jerne 1977). At present, some intermediate hypotheses, in which extra-variability is partly due to somatic mutation and partly due to higher germ line mutation rate at the hypervariable region than at the framework region, seem to have supporters (Weigert and Riblet

1977, Weigert et al. 1978, Valbuena et al. 1978). If somatic mutation is responsible to all or part of extra-variability at the hypervariable region, the formula (6.8) is modified into the following, by letting v_{som} be the somatic mutation rate per amino acid site per generation.

$$C_{som} = \frac{(1 - v_{som})^2}{1 + (A + Bn_{sf})v_{aa}} \qquad (6.9)$$

in which C_{som} is the identity coefficient under the somatic mutation hypothesis.

Equation (6.8) indicates that a linear relationship is expected between the subfamily size (n_{sf}) and the reciprocal of the identity coefficient ($1/C_w$). Note that $1/C_w$ is analogous to the "effective number of alleles" used by Kimura and Crow (1964) to measure gene diversity at a locus in a population. This has been quite useful for analysing the data at enzyme loci. Thus I compute the reciprocal of the average identity coefficient of amino acid for various subgroups compiled in Kabat et al. (1976). The following sequences in Table 6.3 are used. As seen from the table, the number of sequences in some subfamilies is small; however the identity coefficient is fortunately a rather insensitive quantity to the sample size (Nei and Roychoudhury 1974).

In the table, the average identity coefficients at the framework region and at the hypervariable region are also shown. As pointed out before, the whole variable region is known to consist of V and J parts, and genes of these two parts form two different gene families (e. g. Tonegawa et al. 1978, Sakano et al. 1979, Seidman et al. 1979). Therefore only the V part was used for the analysis of the light chain (amino acid site no. 0 ~ 94 by Kabat et al.'s (1976) numbering system). As to the heavy chain, detailed information is not available at the moment and the whole variable region sequence was used. Under the germ line theory, the reciprocal of the identity coefficient at the framework regions (X) and at the hypervariable regions (Y) may be expressed as, from the formula (6.8),

$$X = \frac{1}{C_w FW} = 1 + (A + Bn_{sf})v_{aa}^{FW} \qquad (6.10)$$

and

Table 6.3 Protein families, subfamilies and sequences used in the analyses.

Protein Family	Sub-family	Symbol in Fig. 6.3	Sequences	Average identity coefficient C^{FW}	C^{HV}
human κ	I	a	AG, DEE, HAU, OU, ROY, BEL, EU, CAR, AU, SCW, BI, BJ26, BJ19, GAL(1), REI, NI, KA	0.8761	0.6200
	II	b	TEW, MIL, CUM	0.9675	0.7857
	III	c	TI, FR4, RAD	0.9492	0.8667
mouse κ		d	MOPC70, MOPC321, TEPC124, MOPC63	0.9462	0.8686
rabbit κ	I	e	4135, BS-1, BS-5, K-25	0.9304	0.7963
human λ	I	f	HA, NEW, VOR, NEWM	0.8611	0.6013
	II	g	VIL, NEI, BOH, TRO	0.8879	0.7659
	III	h	X, BAU, KERN, DEL	0.8621	0.5885
mouse λ		i	MOPC104E, RPC20, J558, XS104, SAPC178, SAPC166, Y5606, Y5444, HOPC2020, MOPC315	0.9714	0.9256
human h	II	j	DAW, OU, COR, HE, NEWM	0.7433	0.4613
	III	k	NIE, TEI, WAS, JON, LAY, POM, GAL, ZAP, TUR, TRO, GA	0.7933	0.4720
mouse h	III	ℓ	MOPC173, HOPC8, TEPC15, S107, MCPC603, MOPC167, MOPC511, W3207	0.8684	0.8062
rabbit h		m	BS-5, K-25, 2690, AA1 POOL, BS-1	0.8625	0.6885

In the table, κ, λ and h denote kappa, lambda and heavy chains. The chains of which the majority of sites are sequenced are used. Subgrouping of the protein family by Kabat et al. (9) is adopted, except mouse κ in which four sequences above are chosen as one group for their similarity. All available sequences of mouse λ or rabbit h are used as one subfamily due to their restricted amino acid diversity. The superscripts FW and HV to C in the last two columns denote the framework and the hypervariable regions respectively.

$$Y_{ger} = \frac{1}{C_w^{HV}} = 1 + (A + Bn_{sf})v_{aa}^{HV} \tag{6.11}$$

in which the superscripts FW and HV denote the framework and the hypervariable regions respectively and the subscript "ger" denotes the germ line theory. If the somatic mutation contributes to lowering C_w^{HV}, $v_{som} > 0$ in equation (6.9), therefore, Y becomes,

$$Y_{som} = \frac{1 + (A + Bn_{sf})v_{aa}^{IN}}{(1 - v_{som})^2} \tag{6.12}$$

in which v^{IN} is the germ line mutation rate per amino acid site at the hypervariable region under the somatic mutation theory and the subscript "som" denotes the somatic mutation theory. One may assume

$$v_{aa}^{FW} \leq v_{aa}^{IN} < v_{aa}^{HV}. \tag{6.13}$$

Note that the simplest form of the somatic mutation theory assumes $v_{aa}^{FW} = v_{aa}^{IN}$, since all extra variability at the hypervariable region is due to somatic mutation under this hypothesis. Under the intermediate hypothesis in which extra variability is partly due to somatic mutation and partly due to high mutation rate in germ cells, v_{aa}^{IN} would take the value between v_{aa}^{FW} and v_{aa}^{HV} as in formula (6.13).

From the equations (6.10), (6.11) and (6.12), we obtain the following formulas

$$Y_{ger} = 1 + \frac{v_{aa}^{HV}}{v_{aa}^{FW}} (X - 1) \tag{6.14}$$

and

$$Y_{som} = \frac{1}{(1 - v_{som})^2} \left\{ 1 + \frac{v_{aa}^{IN}}{v_{aa}^{FW}} (X - 1) \right\}. \tag{6.15}$$

Figure 6.3 shows the relationship between $X = 1/C_w^{FW}$ and $Y = 1/C_w^{HV}$ based on actual data of 13 subfamilies listed in Tab. 6.3. The regression line as shown in the figure is

$$Y = 3.61X - 2.62. \tag{6.16}$$

85

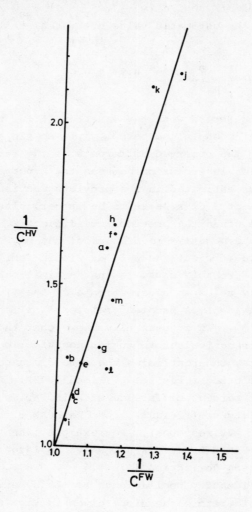

Figure 6.3 Relationship between X = 1/C^FW and Y = 1/C^HV. Data are shown by dots as identified by the symbol given in Table 6.3. Straight line is the regression line; Y = 3.61X - 2.62.

The slope of the regression line, 3.61 ± 0.46 gives the estimate of
the ratio of the germ line mutation rate at the hypervariable region
to that at the framework region (v_{aa}^{HV}/v_{aa}^{FW} or $v_{aa}^{IN}/\{v_{aa}^{FW}(1$
$- v_{som})^2\}$). Also, the estimated value of Y at X = 1 is,

$$\hat{Y}_{X=1} = 0.99 \pm 0.07. \qquad\qquad (6.17)$$

These results are in accord with the equation (6.14) rather than with
the equation (6.15) as explained below. i) The slope of 3.61 can
roughly account for the observed difference of the rates of accumulation of amino acids in evolution between the hypervariable and the
framework regions as estimated in the previous section. However, the
regression coefficient 3.61 appears to be larger, although not statistically significant, than the previous coefficient 2.59 (see page 78)
which is based on the negative logarithms of the between-protein
families identity coefficients as well as within-family identity coefficients. The most probable cause is that estimation by negative
logarithms is likely to be an underestimate of the true divergence
when the comparison involves remotely related families and the identity coefficient is low. It is generally known that the simple Poisson
correction, which is equivalent to taking negative logarithms of i-
dentity coefficients, underestimates the true distance between the
homologous amino acid sequences (e. g. Fitch 1973). Dayhoff (1972)
proposed to convert percent difference of amino acids, which is equal
to 100 × (1 - identity coefficient), into her PAM distance (accepted
point mutations per 100 residues). For example, when the percent
difference is 50%, PAM distance is 83, which is larger than the estimate using the Poisson correction ($-\ln 0.5 = 0.69$). Many values of
the between-family identity coefficients at the hypervariable region
are less than 0.5, therefore the slope given in Fig. 6.2 is likely to
be underestimated.

ii) $\hat{Y}_{X=1}$ of 0.99 is just as predicted by the germ line theory
(formula 6.14). It should be larger than one if somatic mutation
is responsible for hypervariability (formula 6.15). At the moment,
with available data, $\hat{Y}_{X=1}$ has a rather large standard error (0.07),
and $\hat{Y}_{X=1} > 1$ may not be completely denied, however the results
strongly suggest the correctness of the germ line theory.

Additional analysis was performed by using the total protein
family such as human κ or human heavy chains rather than subfamilies
to calculate identity coefficients. In this case, eight pairs of

observed X and Y (see the first column of Table 6.3) are available and an almost identical relationship was obtained: $Y = 3.56X - 2.63$. Here the two regression lines are almost independent estimates even if the same data are used, because the identity coefficient of a protein family often comes from the amino acid identity <u>between</u> the sub-families as well as within a subfamily. At any rate, the agreement of the two regression lines strengthens the conclusion. While the mouse genome contains only one or two genes for λ chain variable region, 8 different λ chains are found among monoclonal immunoglobulins excreted by myelomas of BALB/c origin. The above has often been cited as an evidence favoring the somatic mutation hypothesis (Weigert and Riblet 1977). However, it should be noted that an inbred line such as BALB/c maintained in numerous laboratories constitutes a huge population today. Therefore polymorphisms are expected for complex gene systems like the immunoglobuline gene family (Smith 1977). It is very likely that different sequences of λ chains observed are specified by alleles of the λ chain variable region locus or loci.

A similar argument applies to the other protein families of immuno-globulin. Based on the non-random association of different framework regions and the large number of sequences of κ chain of BALB/c origin as compared with the estimated gene number, Kabat et al. (1979) suggest the possibility of "minigenes," such that the total number of sequences may be increased by their combination. Although this idea is essentially correct to the extent of the combinatorial use of V genes and J segments, there is no evidence that V genes themselves are split into pieces in germ cells, and the observed large number of sequences is likely to be the result of polymorphism within the inbred line.

6.3 Variance and covariance of identity coefficients

As shown above, the pattern of differentiation of immunoglobulins agrees well with the prediction of the germ line theory with respect to the mean coefficient of amino acid identity. It becomes even more convincing if the variance and covariance of identity coefficients also fit the prediction of the germ line theory. Since we measure identity at each amino acid site, the variance of identity probabili-ties <u>among the sites</u> is analogous to the variance of heterozygosity (and gene identity) among the enzyme loci in population genetics (Nei 1975). In the present case, however, the situation is more complicated, since amino acid sites in the variable region are tightly linked and are not independent. However, as I shall show by using the theory developed in Chapter 5, the agreement between the observed and the predicted variances is fairly good.

Now, the variance of amino acid identity among sites, i. e. the variance of $C_{w2,r}$ or that of $C_{b,r}$ for an appropriate set of r (amino acid sites) of a given type of immunoglobulins corresponds to the among-population component of the variance of the probability of gene identity. Similarly, the covariance between $C_{w2,r}$ and $C_{b,r}$ of an immunoglobulin type corresponds to the among-population component of covariance between the within-population and the between-population identity coefficients. Their theoretical predictions are expressed by the equations (5.46) $\sim$ (5.51) in the previous chapter. The variance and the covariance are calculated in the usual way separately for the framework and for the hypervariable regions.

$$\sigma_{bb}^2(FW) = \left\{ \sum_r C_{b,r}^2 - s_{fw}(\bar{C}_b^{FW})^2 \right\} \Big/ (s_{fw} - 1) \qquad (6.18)$$

and

$$\sigma_{2b}(FW) = \left\{ \sum_r C_{w2,r} C_{b,r} - s_{fw}\bar{C}_{w2}^{FW} \bar{C}_b^{FW} \right\} \Big/ (s_{fw} - 1), \quad (6.19)$$

where s_{fw} is the number of amino acid sites in the framework region for which sequence data are available. The variance and covariance for the hypervariable region, $\sigma_{bb}^2(HV)$ and $\sigma_{2b}(HV)$, may be similarly calculated. The same set of sequence data as used in the previous section are used.

In order to compare observed values with theoretical predictions, it is necessary to make corrections for the sampling variance to obtain estimates of variance and covariance. The situation is again analogous to the problem of correcting for the sampling variance in the estimation of probabilities of gene identity at enzyme loci in natural populations. For the present purpose, I use Nei and Roychoudhury's (1974) method. Actually, equation (6.1) does not give an unbiased estimate when the sample size is small. However this estimate is superior to the unbiased one because expected squared deviation from the true value is usually smaller in the former than in the latter (Nei and Roychoudhury 1974). By calculating the correction factor, the effect of this bias is also taken into account. The correction factor is usually small and the correction does not sub-stantially alter the results.

The variance and covariance thus obtained are compared to the theoretical predictions based on the model described previously. The theoretical values of the mean, variance and covariance of the identity coefficient at t = 0 are obtained by using equations (5.52-5.63) in equilibrium. Starting from these values, the necessary moments for identity coefficients between the isolated populations can be obtained. By using the equations (5.7) and (5.66)-(5.77), the moments involving the identity coefficient between populations are generated from one generation to the next and variance and covariance are calculated. Exact values of parameters are not known, but I shall tentatively use the following; $2N_e = 5 \times 10^4$ based on theoretical consideration of the amount of genetic variability at enzyme loci (Kimura and Ohta 1971), $\beta' = 10^{-4}$ which is a conservative estimate based on the observed frequency of recombinants (Weigert and Riblet 1977), $n_g = 100$ which is again a conservative estimate from the recent studies of gene counting method (Seidman et al. 1978, Weigert et al. 1978), $\alpha = \alpha' = 5 \times 10^{-8}$ which is the value when $\gamma = \gamma' = 10^{-4}$ with mean gene number of shift of 10% of n_g, since $\alpha = m\gamma/n_g^2$ and $\alpha' = m'\gamma'/n_g$, the mutation rate per amino acid site per generation at the framework region $v^{FW} = 5 \times 10^{-9}$ based on the rate of amino acid substitution (see Table 6.2) and finally the mutation rate at the hypervariable region $v^{HV} = 3.5 \times v^{FW} = 1.75 \times 10^{-8}$ based on the regression analysis in the previous section. Then the expected average identity coefficients at the framework region and the hypervariable region become respectively, 0.79 and 0.52, and both agree to the observed values.

The application of our theory to the observed variance and covariance of amino acid identities has some shortcomings. Firstly the existence of subfamilies as discussed in the previous section indicates that the genes are clustered and not randomly arranged on the chromosome. Secondly, as will be discussed in the next chapter, amino acid sites are not freely recombined and therefore linkage disequilibrium among the amino acid sites is not negligible. The above two points are not independent of each other and would have the effect of reducing the variance of identity coefficients among the amino acid sites. Thirdly, some of the parameters assigned above are erroneous. However, as long as the relationship between the mean and the variance or covariance of identity coefficients is concerned, wrong assignment of some of the parameters has very little effect. The magnitude of variance appears to depend heavily on the identity coefficient at equilibrium, $\hat{C}_{w2}$, and not on individual parameter. In fact, if I use the set of parameters in my previous report (Ohta 1978c), i. e., $2N_e$ = 10^4, β' = 10^{-3}, n_g = 500, α = 0, α' = 2×10^{-7} and v^{FW} = 5×10^{-9}, almost exactly the same relationship is obtained between C_b and σ_{bb}^2 or σ_{2b}, as in the present set of parameters.

The comparison between the observed and the theoretical variances is given in Figure 6.4. The abscissa is the mean identity coefficient (C_b) and the ordinate is the variance (σ_{bb}^2). The curves represent the theoretical relationship between the two. The solid curve is for the framework region and the broken curve is for the hypervariable region. The observed values obtained from amino acid sequence data are shown by the dots. The solid dots represent values for the framework region, and the open circles, the hypervariable region. Each dot may be identified by a letter or a number as given in Table 6.1. As can be seen from the figure, the agreement between the theoretical predictions and the observed values is reasonably good even if the present model may be too simple.

Figure 6.5 shows the comparison between the observed and the theoretical covariance by using the same data as before. Again, the solid curve is for the framework region and the broken curve is for the hypervariable region. The observed values are shown as solid dots for the framework region, and the open circles for the hypervariable region. The agreement between the observed and the theoretical values is again reasonably good.

The observed variance appears to be somewhat smaller than the predicted value from our simple model. This is probably due to linkage disequilibrium among the amino acid sites. As explained in Chapter 1, a variable region consists of V gene and J segment. If a whole V region behaves as a single block in evolution, the variance of identity coefficients among the amino acid sites should be considerably reduced. Analyses in Chapter 7 indicate that recombination is expected between the genes within a subfamily to some extent, thus a V region is occasionally separated in evolution. Also, the amino acid sites are independent between V and J regions. Yet non-random association of amino acids should be effective in reducing the variance. Further note that, in general, as stated in Chapter 5, the variance becomes large during the differentiation of the gene families when the equilibrium identity coefficient ($\hat{C}_{w2}$) is high, whereas it gets small when the identity coefficient is low. The situation is analogous to the differentiation of selectively neutral enzyme loci.

Natural selection for or against gene diversity as will be discussed in Chapter 8 may have some effects. Here negative selection

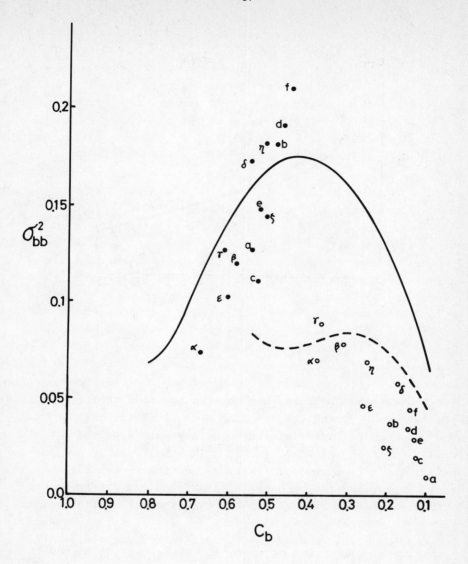

Figure 6.4 Relationship between the mean (C_b) and the variance (σ_{bb}^2) of identity coefficients. Curves
represent theoretical relationship and, the dots,
observed values. The solid curve and dots are for
the framework regions, and the broken curve and
open dots for the hypervariable regions. Symbols
for dots are as given in Table 6.1.

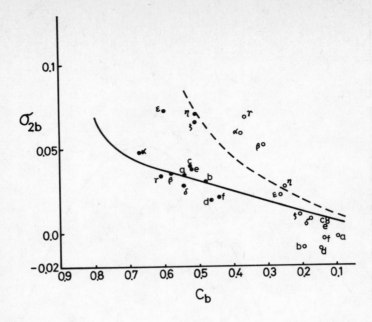

Figure 6.5   Relationship between the mean (C_b)
and the covariance (σ_{2b}) of identity
coefficients. As in Figure 6.4, the
solid curve and dots are the theoretical and observed values of the
framework regions, and the broken curve
and open dots are those of the hypervariable regions.

due to structural constraint of immunoglobulins, as considered to be
responsible for the low mutation rate in the framework region (v^{FW}),
may increase the variance, because the constraint is particularly
rigid in some sites and not so in others. On the other hand positive
selection for antibody diversity in the hypervariable region may
decrease variance if positive selection works uniformly over the
sites in the hypervariable region. This is because selection would
uniformly reduce the identity coefficient of all sites concerned.
    Despite all these difficulties in applying our model to real
data, the results obtained here strongly support the germ line

theory. Both germ-line unequal crossing-over and genetic drift are
random processes, and therefore, even though the mean identity coefficient between the gene families of two species decreases at a
constant rate, the differentiation at individual sites is accompanied
by random fluctuation. The observed variances of such fluctuation
are shown to agree roughly with expectations of the simple model in
the case of the immunoglobulin gene family.

CHAPTER 7

LINKAGE DISEQUILIBRIUM BETWEEN AMINO ACID SITES

IN A MULTIGENE FAMILY

In this chapter, linkage disequilibrium between amino acid sites
(or more generally nucleotide sites) within a gene such as V gene of
immunoglobulin families (see page 6 for organization of immunoglobulin
genes) is investigated. In the following, the amino acid sites are
treated, although the result may be applied to the nucleotide sites.
Linkage disequilibrium is the degree of non-random association of
amino acids between the two sites in a gene family. As discussed in
Chapter 6, the presence of subgroups in an immunoglobulin gene family
indicates non-random association of amino acids in polypeptides. Is
it possible to measure such non-random association quantitatively?
Two-locus problems in population genetics (Hill and Robertson 1968,
Kimura and Ohta 1971b) are analogous and may provide a useful tool to
the problem. In our case, non-random association is due to the
expansion of some lucky combinations by unequal crossing-over (see
Figure 1.7), whereas linkage disequilibrium is created in a population
by natural selection or by random genetic drift between the two loci.
Following my report (Ohta 1980b), the analyses on linkage disequilibrium between amino acid sites and their application to sequence
variability of immunoglobulins are presented.

7.1 Variance of linkage disequilibrium coefficients

The same model and parameters as in Chapter 3 are used; N_e
= effective population size, n_g = number of genes in a family, β
= rate of equal inter-chromosomal crossing-over, γ and γ' are
respectively the rate of unequal intra- and inter-chromosomal
crossing-overs and m and m' are respectively the mean number of gene
shifts at intra- and inter-chromosomal unequal crossing-overs.
However, instead of taking the gene as a unit, we take the amino acid
site as a unit and let v_{aa} be the mutation rate per amino acid site
per generation as in Chapter 6. We further need to define the
following; let us assume that each gene contains n_s sites as in
Figure 7.1, and fix our attention to a particular pair of sites, A
and B (Figure 7.2). Although the shift at unequal crossing-over

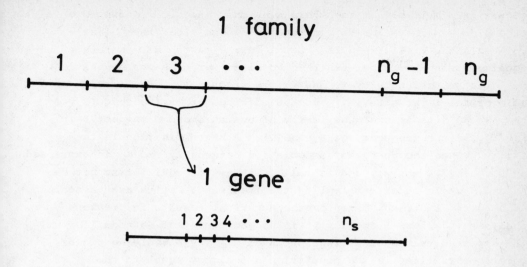

Figure 7.1 Diagram illustrating the fine structure of a multigene family.

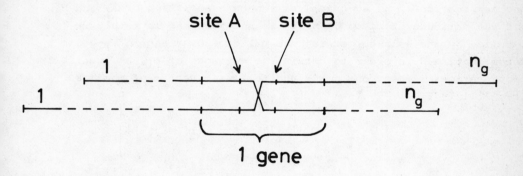

Figure 7.2 Diagram showing recombination between two sites, A and B, at unequal crossing-over.

occurs as a multiple of the whole gene unit, actual crossing-over may take place anywhere inside the gene and we consider recombination between A and B (Figure 7.2). Let ξ be the recombination rate at unequal crossing-overs between the two sites per gene per generation.

Let us designate by x_i and y_j the frequency of the i-th amino acid of the first site (A_i) and that of the j-th amino acid (B_j) of the second site in the gene family of one chromosome respectively. Let f_{ij} be the frequency of a gene with $A_i B_j$ combination in the family. Then the degree of association between A_i and B_j is expressed by $D_{ij} = f_{ij} - x_i y_j$, and is called linkage disequilibrium between A_i and B_j. In order to simplify the treatment, let us assume again $4N_e v_{aa} \ll 1$. Then, based on the numerical results in previous analyses (chapters 3 and 5), it is conjectured that, at equilibrium, the expected values of crossproducts of x_i or y_j involving two or more gene families in the population are almost equal to the corresponding moments involving a single gene family. In the following analyses, I assume that the same principle applies to the two-site model, in which the expectations of various crossproducts are almost equal whether they concern a single gene family or two or more families in the population.

Due to the expansion of certain combinations of A_i and B_j through unequal crossover, linkage disequilibrium increases, whereas it decreases by intra-genic recombination with the rate ξ each generation. Here the expected value of disequilibrium would be zero for any combination ($D_{ij} = 0$ implies random association of A_i and B_j), yet the variance of disequilibrium is not zero and is mainly determined by the balance between unequal crossover and intra-genic recombination. In order to obtain the variance of D_{ij} at equilibrium among various forces, one needs the following vector of moments (cf. Hill 1975).

$$
U = \begin{bmatrix} X \\ Y \\ Z \end{bmatrix} = \begin{bmatrix} E\{(1 - \Sigma x_i^2)(1 - \Sigma y_j^2)\} \\ 4E(\underset{i\,j}{\Sigma\Sigma}\, x_i y_j D_{ij}) \\ 2E(\underset{i\,j}{\Sigma\Sigma}\, D_{ij}^2) \end{bmatrix} \tag{7.1}
$$

where E denotes taking the expectation and summation is for all possible amino acids (1 ∿ 20). In the following, the theory for the change of U from one generation to the next will be given and then the equilibrium properties of U will be presented.

i)   Change of U due to unequal crossing-over

Since the process of unequal crossing-over may be treated by the theory of random genetic drift (see Chapters 2 and 3), the same transition matrix for the moments of the two-locus problems may be used. Here the crucial factor is the rate of increase per generation of an identity coefficient, $C_A = \sum_i x_i^2$, which is the probability of identity of two homologous units taken from the gene family. This rate is shown to be $\alpha = m\gamma/n_g^2$ and is analogous to the rate of increase of homozygosity $(1/2N_e)$ due to random genetic drift in a finite population. U changes into U' through one generation of unequal crossing-over. Let α be much less than unity. Then the transition matrix of U takes the following form ignoring the higher order terms of α (Hill and Robertson 1968, Ohta and Kimura 1969)

$$U' = KU \tag{7.2}$$

with

$$K = \begin{bmatrix} 1 - 2\alpha & \alpha & 0 \\ 0 & 1 - 5\alpha & 2\alpha \\ 2\alpha & 2\alpha & 1 - 3\alpha \end{bmatrix} . \tag{7.3}$$

When inter-chromosomal crossing-over is unequal, the treatment becomes more complicated. However, when $4N_e v_{aa}$ (four times the product of the effective population size and mutation rate per site) is much less than unity, the inter-chromosomal crossing-over process is almost the same as the intra-chromosomal one in terms of the expansion-contraction process of gene units (see Chapter 3). Here inter-chromosomal crossing-over consists of the expansion-contraction process and the gene exchange process between chromosomes. In the following, I shall consider the simple case in which $4N_e v_{aa} \ll 1$. It can be shown that the corresponding change of U to the formulas (7.2) and (7.3) by the contraction-expansion process of inter-chromosomal crossing-over takes the following form.

$$U' = K'U + A' \tag{7.4}$$

where

$$K' = \begin{bmatrix} 1 - 2\alpha'(1 + 4n_g N_e v_{aa}) & \alpha' & 0 \\ 0 & 1 - 5\alpha' & 2\alpha' \\ 2\alpha' & 2\alpha' & 1 - 3\alpha' \end{bmatrix} \qquad (7.5)$$

$$A' = \begin{bmatrix} 8n_g N_e v_{aa}\alpha'(1 - \hat{C}_{w1}) \\ 0 \\ 0 \end{bmatrix} \qquad (7.6)$$

in which α' is equal to $m'\gamma'/n_g^2$ and $\hat{C}_{w1}$ is the identity coefficient at a single site. For details of the derivation of formulas (7.5) and (7.6), see Ohta (1980b). The above formula may be applicable only when $4N_e v_{aa} \ll 1$. When this condition is not satisfied, the expectations of various crossproducts other than X, Y and Z become necessary. As long as one considers amino acid sites as units to examine linkage disequilibrium, such a condition would usually be satisfied.

ii) Change of U due to intra-genic recombination

Disequilibrium is reduced with the rate of intra-genic (between sites) recombination (ξ) each generation. U' changes to U" in one generation according to the formula (e. g. Hill and Robertson 1968)

$$U" = QU' \qquad (7.7)$$

with

$$Q = \begin{bmatrix} 1 & 0 & 0 \\ 0 & 1 - \xi & 0 \\ 0 & 0 & (1 - \xi)^2 \end{bmatrix} . \qquad (7.8)$$

Note that the rate ξ should not include the case of equal crossing-over. The intra-genic recombination effectively takes place when unequal crossover occurs between the sites of paired homologous genes. Unequal crossing-over may be intra-chromosomal or inter-chromosomal. It is assumed here that the arrangement of gene lineages is more or less random along the chromosome, hence the unequal pairing between the homologous genes implies that each unit pairs with a random member of the gene family.

iii) Change of U due to inter-chromosomal (equal) crossing-over

By this process, only the exchange of gene members between the chromosomes is taken into account and neither unequal crossing-overs nor intra-genic recombinations are considered (see Figure 3.2). Let us assume that the gene lineages are randomly arranged along the chromosome and suppose that, by the crossing-over, the gene family is divided into two parts; r and 1 - r where r is assumed to be uniformly distributed. Under these assumptions, it is possible to calculate the amount of change of various quantities as in Chapters 3 and 5. The assumption of random arrangement of gene lineages on the chromosome is not usually satisfied. When the mean number of gene shifts at unequal crossing-over (m or m') is more than 10% of the total gene number in a family, gene arrangement becomes roughly random (see Chapter 4).

As to the first element, X, one needs the theory of the single-site model of the multigene family; the equilibrium and the transitional properties of the identity coefficient investigated in Chapter 3. The first element, X, may be expressed, $X = E\{(1 - C_A)(1 - C_B)\}$ in which C_A and C_B designate the identity coefficient of the first and the second sites respectively. At equilibrium,

$$E(C_A) = E(C_B) = \hat{C}_{W1}$$

$$= \frac{\alpha + \alpha'}{\alpha + \alpha' \frac{1+4n_g N_e v_{aa}}{1+4N_e v_{aa}} + 2v_{aa} + \frac{\beta'}{3} \frac{4N_e v_{aa}}{(1+4N_e v_{aa})}} \qquad (7.9)$$

and the change of X due to inter-chromosomal crossover becomes $\Delta X \approx -2\Delta\hat{C}_{w1}(1 - \hat{C}_{w1})$. It was also shown that the change of $\hat{C}_{w1}$ due to inter-chromosomal exchange is $\Delta\hat{C}_{w1} \approx \frac{\beta'}{3} \frac{4N_e v}{1 + 4N_e v} \hat{C}_{w1}$ at equilibrium (Chapter 3). Therefore

$$\Delta X \approx \frac{2\beta'}{3} \cdot \frac{4N_e v_{aa}}{(1+ 4N_e v_{aa})} \left\{ (1 - \hat{C}_{w1}) - X \right\}$$

$$\approx \frac{8N_e v_{aa}\beta'}{3} \left\{ (1 - \hat{C}_{w1}) - X \right\} \qquad (7.10)$$

when $4N_e v_{aa} \ll 1$.

The change of the second and the third elements of U (Y and Z) due to inter-chromosomal crossing-over may be obtained more directly than that of X, by taking the expectation of the change of Y and Z again assuming the uniform distribution for r. They can be shown to be

$$\Delta Y = -\Delta Z \approx -\frac{4\beta'}{30} E\{(\hat{C}_{w1} - \hat{C}_{w2})^2\} \tag{7.11}$$

in which $\hat{C}_{w2}$ is the probability of identity of two homologous units taken from <u>different</u> chromosomes and $\hat{C}_{w2} = \hat{C}_{w1}/(1 + 4N_e v_{aa})$ at equilibrium (see Chapter 3). When $4N_e v_{aa} \ll 1$, $\hat{C}_{w1} \approx \hat{C}_{w2}$ and the above formula (7.11) becomes zero. In the following part of this section, I shall use $\hat{C}_w$ for $\hat{C}_{w1}$ or $\hat{C}_{w2}$. Thus U" changes to U''' by inter-chromosomal crossing-over as follows

$$U''' = RU'' + B \tag{7.12}$$

in which

$$R = \begin{bmatrix} 1 - \dfrac{8N_e v_{aa}\beta'}{3} & 0 & 0 \\[2ex] 0 & 1 & 0 \\[2ex] 0 & 0 & 1 \end{bmatrix} \tag{7.13}$$

and

$$B = \beta' \begin{bmatrix} \dfrac{8N_e v_{aa}}{3}(1 - \hat{C}_w) \\[2ex] 0 \\[2ex] 0 \end{bmatrix}. \tag{7.14}$$

As long as the moments concerning a single gene family are almost equal to their corresponding moments concerning two or more gene families in the population as assumed in the present analyses, the moments under consideration would not be influenced by random sampling of chromosomes. However note that the sampling factor (N_e) comes into the formulas (7.13) and (7.14) through its effect on the individual site.

iv) Change of U due to mutation

Since all mutations are assumed to be unique, every element of U reduces to $(1 - v_{aa})^4$ of the previous value each generation by mutation (e. g. Kimura and Crow 1964). Not only the reduction of the moments but also their increments due to new mutations need to be considered (Ohta and Kimura 1971, Hill 1975). They can be shown to become

$$
F = \begin{bmatrix} \Delta X_m \\ \Delta Y_m \\ \Delta Z_m \end{bmatrix} = \begin{bmatrix} 4v_{aa}(1 - \hat{C}_w) \\ 0 \\ \dfrac{4v_{aa}(1 - \hat{C}_w)}{n_g^2} \end{bmatrix}.
$$

Therefore, U''' changes to U'''' due to mutation by the following formula,

$$
U'''' = (1 - v_{aa})^4 U''' + F. \tag{7.15}
$$

The total change of U per generation may be obtained by combining the above equations (7.2) $\sim$ (7.15),

$$
U_{t+1} = (1 - v_{aa})^4 KK'QRU_t + A' + B + F \tag{7.16}
$$

in which t is the time measured in generations and it is assumed that all parameters α, $n_g\alpha'$, β', ξ and $4N_e v_{aa}$ are much smaller than unity. At equilibrium, $U_{t+1} = U_t = \hat{U}$ and the moments may be readily obtained by solving the following equation for $\hat{U}$.

$$
\hat{U} = (1 - v_{aa})^4 G\hat{U} + A' + B + F \tag{7.17}
$$

where

$$G = KK'QR$$

$$\approx \begin{bmatrix} 1-2\alpha-2\alpha'(1+4n_g N_e v_{aa}) - \dfrac{8N_e v_{aa}\beta'}{3} & \alpha+\alpha' & 0 \\ 0 & 1-5(\alpha+\alpha')-\xi & 2(\alpha+\alpha') \\ 2(\alpha+\alpha') & 2(\alpha+\alpha') & 1-3(\alpha+\alpha')-2\xi \end{bmatrix}.$$

$$(7.18)$$

7.2 Application to sequence variability of immunoglobulins

The immunoglobulin gene family is again a most interesting example for application of the theory. By using the compilation of variable region sequences by Kabat et al. (1976), I shall apply the present theory to the statistical analyses on sequence variability of the variable region of immunoglobulins.

Non-random association between amino acids of the two sites of a chain results in the loss of genetic variability of the family as a whole. When the association is not random, the disequilibrium coefficient (D) is not zero and the first element of $\hat{U}$, $X = E\{(1 - C_A)(1 - C_B)\}$, becomes larger than the product of the expected values of $(1 - C_A)$ and $(1 - C_B)$ at the individual site; $X > (1 - \hat{C}_w)^2$. Here, let us define a quantity, Δ, as the excess probability of simultaneous identity over that expected from random combination of the identity at the two sites. Δ may be expressed as; $\Delta = E(C_A C_B) - E(C_A) \times E(C_B) = X - (1 - \hat{C}_w)^2$.

By applying the theory to the actual data of immunoglobulins, let us consider the following comparison of homologous amino acid sequences. If the two sequences compared have the identical amino acid at the A site, we score one ($C_A = 1$). If the homologous amino acids are different, we score zero ($C_A = 0$). If we average the score over all sites compared, the average value is equivalent to the identity coefficient, which is the probability of identity of two homologous amino acids. As to the reported sequences of the variable region of immunoglobulins, we shall examine all pairwise comparisons for a given type of immunoglobulin of a species such as human κ. Let us designate the average identity coefficient over all pairwise comparisons as $\bar{I}$ and the variance among the comparisons as V_I. The expectations of $\bar{I}$ and V_I can be shown to be approximately, by letting n_s be the number of sites compared,

$$E(\bar{I}) \;=\; \hat{C}_w \tag{7.19}$$

and

$$E(V_I) \;=\; \frac{n_s - 1}{n_s}\, \overset{\sim}{\Delta} + \frac{\hat{C}_w(1 - \hat{C}_w)}{n_s} \;\approx\; \overset{\sim}{\Delta} \tag{7.20}$$

in which $\overset{\sim}{\Delta}$ is the average value of Δ for all pairs of two sites compared for identity in the variable region, and the approximate equality holds when n_s is sufficiently large. This relationship implies that simultaneous identity of amino acids at the two sites is expected to be larger by the amount of Δ than random combination of amino acids predicts.

By using the theory in the previous section, it is possible to get a rough idea of the magnitude of Δ. For this purpose, we use "squared standard linkage deviation (σ_d^2)" which is defined as the ratio of Z (third element of U) to twice X (Ohta and Kimura 1969, 1971, Hill 1975). From the equation (7.17), σ_d^2 is shown to become roughly

$$\sigma_d^2 \;=\; \frac{Z}{2X} \;\approx\; \frac{1}{3+4v'+2\xi'-4/(5+4v'+\xi')} \tag{7.21}$$

where $v' = v_{aa}/(\alpha + \alpha')$ and $\xi' = \xi/(\alpha + \alpha')$. Then Δ may be expressed by the following approximate formula,

$$\Delta \;=\; \frac{4\sigma_d^2}{(2+4v'+b')(5+4v'+\xi')}\, X \;\approx\; \frac{4\sigma_d^2(1-\hat{C}_w)^2}{(2+4v'+b')(5+4v'+\xi')} \tag{7.22}$$

where $b' = 8N_e v_{aa}(\beta' + n_g\alpha')/\{3(\alpha + \alpha')\}$. For details, see Ohta (1980b). When $\hat{C}_w$ is close to 1, the last approximate equality does not hold. Let us call the quantity, $\Delta/(1 - \hat{C}_w)^2$, "standard identity excess".

I have calculated the mean ($\bar{I}$) and variance (V_I) for various groups of variable region sequences of immunoglobulins. The same set of sequence data as in Chapter 6 is used which are compiled by Kabat et al. (1976). Recently it is reported that the whole variable region of the light chain actually consists of V (variable) region and J (junction) region and apparently the regions constitute

different gene families (see Chapter 1). Thus I exclude from my
analyses the J region, which has only about 15 amino acid sites at
the junction part with the constant region. By using the numbering
system of Kabat et al., the V region is 0 ∿ 94th amino acid sites.
As to the heavy chain, such detailed information has not yet been
made available and I use the entire variable region sequence in my
analyses.

$\bar{I}$ and V_I are given in Table 7.1 for various groups of immunoglobulins. $V_I/(1 - \bar{I})^2$ is also given, the expected value of which is

Table 7.1 Mean ($\bar{I}$), variance (V_I) and the ratio ($V_I/(1-\bar{I})^2$)
of amino acid identity for pairwise comparisons
of variable region sequences of immunoglobulins.
The standard identity excess may be estimated by
the ratio.

	$\bar{I}$	V_I	$V_I/(1-\bar{I})^2$	number of pairs
Human κ	0.6900	0.01255	0.1306	276
Mouse κ	0.6008	0.01815	0.1139	45
Rabbit κ	0.7277	0.00986	0.1330	136
Human λ	0.6230	0.00930	0.0654	105
Human H	0.5570	0.01773	0.0903	153
Mouse H	0.7240	0.04479	0.5880	45
Rabbit H	0.7896	0.00617	0.1394	10
Human κ subgroup I	0.7776	0.00265	0.0536	136
Huamn λ subgroup I	0.7085	0.00216	0.0254	6
Human H subgroup III	0.7061	0.00361	0.0418	55

standard identity excess. From the equations (7.21) and (7.22), the
standard identity excess is approximately $4/\{(2+4v'+b')(5+4v'+\xi')$
$(3+4v'+2\xi'-4/(5+4v'+\xi'))\}$, which becomes greatest ($4/22 = 0.182$)
when v', b', $\xi' \ll 1$, or when $\alpha + \alpha'$ is much larger than v_{aa}, ξ and

$N_e v_{aa} \beta'$. As in Chapter 6, if $2N_e = 5 \times 10^4$, $\beta' = 10^{-4}$, $n_g = 100$, $\alpha = \alpha' = 5 \times 10^{-8}$ with $\gamma = \gamma' = 10^{-4}$ and $m = m' = n_g/10 = 10$, and $v_{aa} \approx 10^{-8}$, the observed identity coefficients agree with the theoretical prediction of the model. Then $v' = 0.1$ and $b' \approx 1/3$. If $\xi' << 1$ implying that $\alpha + \alpha' >> \xi$, the ratio becomes about 0.11 which is roughly the same as the observed values for a group of immunoglobulin sequences corresponding to a gene family of a species. The mouse heavy chain appears to be an exception. Close examination of this family and other species further reveals that the variance may be largely attributed to that between sub-groups within a gene family. In fact, a bimodal distribution of identity coefficients is often found for one family. The bottom three lines of Table 7.1 show that V_I within a subgroup is small. Only the cases with sufficient data are shown. In the case of mouse heavy chain, the entire family is subdivided into three subgroups by Kabat et al. (1976) and the subgroups are more differentiated than in other families. Here it is not very meaningful to get $V_I/(1 - \bar{I})^2$ within a subgroup, since $\bar{I}$ is large and close to 1.

As to the subfamilies listed in the bottom three lines of Table 7.1, the variance is much smaller than that of the whole families. Here the variance is not significantly larger than the variance due to a finite number of sites, which is approximately $\bar{I}(1 - \bar{I})/n_s$. Thus fitting the parameters of our model is not very meaningful. However, it is expected that ξ is at least as large as $\alpha + \alpha'$ (= 10^{-7}) and that recombination takes place between genes of one subgroup. Thus, from the present result, I suggest that the subgroups compiled by Kabat et al. (1976) correspond to the gene family as considered here and that the whole group such as human κ or mouse κ consists of several subfamilies between which recombination would be rather limited. However, no evidence is available on the existence of many sub-subgroups as suggested by immunologists (Weigert and Riblet 1977). In other words, recombination does not seem to be restricted to occur between the genes belonging to a very small sub-subgroup which contains only several gene members but appears to take place between any genes in one subgroup.

The analyses presented here are not exact in many respects. First, I did not take into account the effect of sampling sequences from a large population such as human and mouse species. Second, the difference in mutation rates among the sites was not considered in the analyses. The difference is particularly evident if the

hypervariable and the framework regions are compared. As long as
linkage disequilibrium is concerned, however, the difference of
mutation rate would have a minor effect. Thirdly, genetic correlation
with chromosomal distance within the family was not considered. At
the moment, the arrangement of genes in a family is not known,
however the existence of subgroups suggests that each group is
clustered on the chromosome so that genetic correlation would
decrease with chromosomal distance. It is likely that each subfamily
is clustered on the chromosome, however it is still possible that, in
one subfamily, genes are more or less randomly arranged on the
chromosome. It is also expected that, although very rarely, the
unequal crossing-over takes place between the genes of different
subgroups. By chance, a single subfamily may also differentiate into
two subfamilies and thus the total family evolves. An important
aspect is that this process is essentially random. Definition of sub-
subfamilies and assignment of a fixed number of genes to each sub-
subfamily in every genome of a species (e. g. Weigert and Riblet 1977)
seem to be highly arbitrary.

On the other hand unequal crossing-over and gene conversion are
suggested as factors which generate antibody diversity during
ontogeny (Seidman et al. 1978). This is because they create new
associations of amino acids in one sequence. They tend to decrease
linkage disequilibrium during somatic differentiation but this effect
would be small in quantitative analyses such as the present study.

CHAPTER 8

PROSPECTS FOR FUTURE RESEARCH

Theoretical models presented in the previous chapters are based on overly simplified assumptions, such as constant gene family size and no effect of natural selection on the gene diversity contained in the family. It would be preferable to remove such assumptions one by one and to build more realistic models. Many such problems are not solved at the moment and left for future analyses. In this chapter, various unsolved problems are presented. Also included are the possible applications of the theory other than those discussed in the previous chapters.

8.1 Gene conversion and other mechanisms to increase homogeneity

In the previous chapters, unequal crossing-over is considered to be the main cause for maintaining homogeneity among the gene members in a multigene family and investigated in detail. Although, as it is pointed in the first chapter (page 12), unequal crossing-over has the strongest experimental support compared to the other mechanisms considered, it is possible that there are some other processes involved in making gene members homogeneous. Among various possibilities, gene conversion is worthy of consideration. In particular the observation that there are some repeated gene families distributed on non-homologous chromosomes (Glover et al. 1975) suggests the possibility that gene conversion also plays some role on keeping homogeneity of the gene members of the family. It may also be important for the transmission genetics of extra-chromosomal genes such as mitochondrial or chloroplast genes (Birky 1978), as well as for immunoglobulin gene families (Gally and Edelman 1972). At any rate, when multiple copies of genes are present in a germ cell, it seems that the gene conversion should be seriously considered as a factor which might contribute to the homogeneity of multiple gene families.

If gene correction takes place between random members of the gene family in a genome or in a cell as in the democratic gene conversion model of Gally and Edelman (1972), the process is much simplified and can be treated similarly to unequal crossing-over (Ohta 1977). In fact, if a random member corrects another random member of a gene family, the former type increases by one unit and the latter type decreases by one unit, and as a result, one cycle of duplication and deletion (see page 17, Table 2.1) occurs. It does not matter whether the length of the DNA segment converted is long or short, as long as the correction takes place between the homologous segments. However, when it is short, one should consider small units such as amino acid sites or nucleotide sites. If gene rectification occurs between the gene members on different (homologous) chromosomes at meiosis, one conversion is analogous to one cycle of inter-chromosomal unequal crossing-over (see page 30, Table 3.2). Of course, only expansion and deletion of a unit occur without accompanying additional gene exchange between the chromosomes. At any rate, the gene conversion may be treated by the same diffusion process as unequal crossing-over.

It is not known whether or not gene conversion takes place between members of a gene family far apart on the chromosome or on non-homologous chromosomes. If it does, it certainly works for maintaining homogeneity of separated gene families as observed by Tartof and David (1976) and by Finnegan et al. (1978), although occasional transfer of genetic material may be more important for such cases. It is also noted that identical gene families on non-homologous chromosomes are often telomeric, i. e. restricted to the terminal region of chromosomes. In such cases, meiotic recombination would be allowed to occur between such non-homologous chromosomes. Saltatory replication (Southern 1970, Walker 1971, Davidson and Britten 1971) or rolling circle of a gene (Callan 1967) should essentially give the same effect, if such processes only occasionally take place and if a master gene is not fixed but chosen randomly, in the sense that they lead to duplication of some genes at the expense of the others.

8.2 Non-constant gene family size

The assumption that the gene family size n_g stays constant over an evolutionary time scale is unrealistic and, in fact, many observations suggest that it varies considerably from time to time. A most extreme example is the 5S ribosomal RNA gene families in closely related species of Xenopus; X. mulleri has about 9,000 5S rRNA genes while X. laevis has about 24,000 genes (Brown and Sugimoto, 1973). Also, as discussed in Chapter 6, the lambda chain gene family of mouse has much contracted since its divergence from the other families. The rapid differentiation of satellite DNA in some species (Henning and Walker 1970) may be another example in which the size of a family contracted and, in fact, the family may have been totally replaced

with some others. In general, however, functionally important gene
families seem to have been kept fairly stable in their size.

Crow and Kimura (1970, pp. 294-296) considered a model in which
natural selection controls the gene family size such that the fitness
is maximum for an intermediate family size and decreases with the
square of the deviation from this optimum amount. In this model, a
balance will be reached between the generation of length heterogeneity
by unequal crossing-over and the selective elimination of more extreme
variant chromosomes. If it is assumed that the difference of the
family size between the new recombinant family and the parental one is
small, approximately normally distributed, and independent of the
parental family size, then the equilibrium distribution of family size
can be shown to be normally distributed. Here the mean ($\bar{n}_g$) and the
variance (V_{n_g}) of the distribution may be expressed by the following
equations (Crow and Kimura 1970).

$$\bar{n}_g = n_{opt} \tag{8.1}$$

and

$$V_{n_g} = \sqrt{\frac{\gamma m^2}{2s}} \tag{8.2}$$

in which n_{opt} is the optimum n_g, γ is the rate of unequal (intra-
chromosomal) crossing-over, m is the mean number of genes expanded or
deleted, and s is the selection coefficient such that the fitness of
a family with size n_g is $1 - s(n_g - n_{opt})^2$. If one considers inter-
chromosomal unequal crossing-over, γ and m may be replaced by γ' and
m'. It is likely however, that the difference between old and new
family sizes is dependent upon the gene family size. Then the treatment becomes more complicated and awaits future investigation. At
any rate, when gene family size is under control of natural selection
and more or less stable, the average family size may be used for
theoretical investigations on genetic variability as in the previous
chapters. In the future, an "effective gene family size" may be
introduced just like the effective population size so conveniently
used in the population genetics theory.

In connection with gene family size, a very interesting
phenomenon needs attention. The bobbed locus in _Drosophila_ has been
known as an X chromosome linked recessive mutant which causes abnormal

110

bristles, but is now known to code for 18S and 28S ribosomal RNA
(e. g. Tartof 1975). In 1971, Ritossa et al. reported that rDNA
magnification at the bobbed loci is characterized by a stepwise
accumulation and suggested that such a process is caused by unequal
crossing-over within the bobbed loci. Clayton and Robertson (1957)
and Frankham et al. (1978) studied the same problem from the standpoint of quantitative genetics and found, in their selection
experiment on abdominal bristle number, that an abrupt increase of
genetic variation occurs in some selected lines of low bristle number.
The latter authors have attributed this phenomenon to unequal
crossing-over in the ribosomal RNA gene family. Bristle number is
one of the best studied cases of quantitative variation and Frankham
et al. (1978) emphasized the possible importance of unequal crossing-
over on quantitative characters in general, since the functioning of
rRNA loci is crucial to the expression of all protein specifying
loci. Wright (1977, page 374) also discusses the possible significance of repeated genes on quantitative variability, by combining the
observation of Britten and Kohne (1968) with Mather's hypothesis of
oligogenes and polygenes (1949).

8.3 Accumulation of non-functional genes

Since there are multiple copies of genes in a genome, mutations
necessarily accumulate in the gene members of the family. It is now
believed that deleterious mutations are much more frequent than
beneficial mutations particularly at the molecular level (e. g.
Kimura and Ohta 1974), and the accumulation of deleterious or non-
functional mutations in the family is unavoidable to some extent. An
extremely interesting example is the presence of "pseudogenes" in
the 5S ribosomal RNA gene family of Xenopus laevis reported by
Fedoroff and Brown (1978). They found that functional genes and
non-functional pseudogenes are alternately repeated in this gene
family (see Figure 8.1). Apparently, an original non-functional
mutation was duplicated by unequal crossing-over. It is reasonable
to suppose that, by chance, the non-functional gene and a normal
gene have expanded together as a unit through repeated unequal
crossing-overs and, as a result, a pseudogene and a normal gene are
alternately arranged on the chromosome. In general, however, detailed
information of genetic structure of multigene families is not a-
vailable, particularly concerning the presence of non-functional
genes.

From the theoretical standpoint, the mechanism of preventing

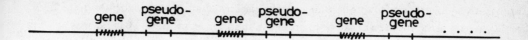

Figure 8.1   A model of the structure of 5S ribosomal RNA
gene family of <u>Xenopus</u> <u>laevis</u>.

degeneration of the gene family should be clarified. Natural selection should operate on the gene family as a whole rather than on
individual mutants in the family (Hood et al. 1975). Perhaps the
simplest model of natural selection to eliminate deleterious mutations
is a truncation type selection in which deleterious mutations are
detected by selection only after they spread to a certain frequency,
y, in the gene family through repeated unequal crossing-overs. Since
the expansion or deletion of mutant genes in the family by unequal
crossing-over is controlled by chance, most of the deleterious genes disappear by chance without selection, just as most selectively neutral
mutations are expected to disappear by chance. Thus, the selective
elimination of inferior individuals from the population, the so-called
mutational genetic load, is much less than the ordinary case where
the deleterious mutants are eliminated by themselves. In fact the
probability of a deleterious mutant of increasing to the frequency y
in the family of n_g gene units is $1/(n_g y)$, so that the mutational
load should be reduced to $1/(n_g y)$ of the conventional prediction.
This statement implies that $n_g y$ mutant genes are eliminated together
by one reproductive death instead of the elimination of mutant genes
one by one.

The above model is too simple and natural selection is likely
to operate on a more quantitative basis. Also, this type of selection
can not be independent of gene family size. In some way, the total
amount of functional gene product should be regulated. A more
detailed quantitative assessment awaits future investigation.

8.4 Natural selection on gene diversity

The analyses of identity coefficients in the previous chapters
are based on the assumption that mutant genes are selectively neutral.
Actually, however, it is possible that natural selection has an

important effect on the genetic variability contained in the gene family. Again, natural selection operates on the gene family as a whole rather than on individual mutants. What is important for survival of the individual is how much genetic diversity is contained in the gene family, at least in the case of immunoglobulin gene families. I have discussed several models in which natural selection works directly on the identity coefficient (Ohta 1978a, b, 1979). Although a general solution for gene diversity with selection has not yet been obtained, I shall summarize my analyses here.

i) Positive selection for gene diversity

Let us assume that selection acts in such a way that the fitness of the family with identity coefficient c_i is

$$W_i = 1 - sc_i \qquad (8.3)$$

where s is a selection coefficient, assumed to be positive. Then we can investigate how the mean values of the identity coefficients (C_{w1} and C_{w2}, see page 23) change by selection. Since C_{w1} is expressed by $C_{w1} = \sum_i c_i p_i$ (see Fig. 3.1), the amount of change of C_{w1} due to selection may be obtained from the frequency change of p_i in the population. The change of p_i by selection in one generation is, as in the ordinary population genetics treatment (Wright 1969),

$$\Delta p_i = p_i' - p_i = \frac{1 - sc_i}{1 - sC_{w1}} p_i - p_i$$

$$= \frac{-s(c_i - C_{w1})p_i}{\bar{W}} \qquad (8.4)$$

where p_i' is the frequency of the ith type of the gene family after selection, and $\bar{W}$ is the mean fitness of the population. Therefore the change of C_{w1} by selection becomes,

$$\Delta C_{w1} = \sum_i c_i (\Delta p_i) = -\sum_i \frac{sc_i(c_i - C_{w1})p_i}{\bar{W}} = \frac{-s\sigma_{w11}^2}{\bar{W}} \qquad (8.5)$$

in which σ_{w11}^2 is the variance of c_i within a population (see page 56, formula 5.40). This formula implies that the mean identity coefficient decreases by an amount $s\sigma_{w11}^2/\bar{W}$ each generation through this type of selection and is analogous to the fundamental theorem of natural selection by Fisher (1930), since this theorem states that the rate of change in fitness at any instant is equal to the variance in fitness at that time. As compared with the effect of mutation and inter-chromosomal crossing-over, the effect of selection may be more pronounced, because the selection coefficient s may be much larger than the mutation rate or the crossing-over rate.

The change due to selection of C_{w2}, the average between-chromosome identity coefficient, may be similarly calculated.

$$\Delta C_{w2} = \sum_i \sum_j c_{ij}\{p_i(\Delta p_j) + p_j(\Delta p_i)\} = \frac{-2s\sigma_{w12}}{\bar{W}} \qquad (8.6)$$

in which σ_{w12} is the covariance between c_i and c_{ij} (see page 56, formula 5.41). My numerical calculations show that σ_{w12} is generally positive, although smaller than σ_{w11}^2, so that C_{w2} also decreases by selection. The results of my simulation experiments are shown below to give an example of the effect of natural selection. For details of the procedure of the simulation, see Ohta (1978a).

Under the present model of selection, genetic differentiation between the two isolated families (populations) is also accelerated. The identity coefficient between the isolated families (denoted as C_b, see page 51) decreases by a constant rate, twice the mutation rate (2v), when there is no selection. With selection, it can be shown that the change of C_b due to selection in one generation becomes,

$$\Delta C_b = \frac{-2s\sigma_{w1b}}{\bar{W}} \qquad (8.7)$$

in which σ_{w1b} is the covariance of the within-chromosome identity coefficient (c_i) and between-isolated families identity coefficient ($e_{i\ell}$) (see page 56, equation 5.43). Since the selection coefficient s is usually much larger than the mutation rate v, and σ_{w1b} is generally positive, the differentiation may be much accelerated.

The selection scheme as defined by formula (8.3), may be modified in various ways. First of all, the fitness function (8.3) takes into account only the additive effect of the identity coefficient of individual chromosomes, but it would be more desirable to be able to incorporate dominance. For the purpose of computing the genetic

Table 8.1 The mean identity coefficient (C_{w1}) and the ratio
of the two measures of identity (C_{w2}/C_{w1}) for the
case with selection (s = 0.5) observed in simulation studies. The expected identity coefficient
and the ratio by approximation formulas (3.22) and
(3.23) under selective neutrality is also given in
parentheses for comparison.

N_e	α	v	β	n_g	C_{w1}	C_{w2}/C_{w1}
12.5	0.08	0.002	0.24	5	0.611 (0.870)	0.587 (0.909)
12.5	0.08	0.002	0.0	5	0.716 (0.952)	0.594 (0.909)
12.5	0.08	0.004	0.24	5	0.601 (0.789)	0.574 (0.833)
12.5	0.08	0.004	0.0	5	0.692 (0.909)	0.519 (0.833)
25	0.02	0.004	0.24	10	0.393 (0.386)	0.564 (0.714)
25	0.02	0.004	0.0	10	0.417 (0.714)	0.425 (0.714)

variation of the gene family of a _zygote_, the identity coefficient
between two different chromosomes should be considered, since in
general $c_{ij} \neq (c_i + c_j)/2$, where c_{ij} is the identity coefficient
between the i-th and the j-th family types. The fitness of the zygote
of the i-th and the j-th family types may then be defined as follows:

$$W_{ij} = 1 - \frac{s}{2}(c_i + c_j + 2c_{ij}) \tag{8.8}$$

In this model, the changes of C_{w1} and C_{w2} by selection in one generation may be shown to be

$$\Delta C_{w1} = \frac{-s}{2\bar{W}}(\sigma_{w11}^2 + 2\sigma_{w12}) \tag{8.9}$$

and

$$\Delta C_{w2} = \frac{-s}{\bar{W}}(\sigma_{w12} + \sigma_{w22}^2) \tag{8.10}$$

where σ_{wll}^2, σ_{wl2} and σ_{w22}^2 are the variances and the covariance of identity coefficients within a population as defined by the formulas (5.40), (5.41) and (5.42) (page 56).

The fitness function may further be modified such that the fitness is now determined by the relative value of the identity coefficient in the population, as in competitive selection. Let us suppose that the fitness of the i-th family type is expressed by the following formula,

$$W_i = 1 - s(c_i - C_{wl}) \qquad \text{for } c_i > C_{wl}$$
$$W_i = 1 \qquad\qquad\qquad \text{for } c_i \leq C_{wl}$$

(8.11)

Here, the i-th family type has selective disadvantage only when c_i is above the population average C_{wl}, or when gene diversity is below an average value. It is difficult to express explicitly the change in C_{wl} or C_{w2} under this selection scheme; although it is clear that the genetic load as defined by the amount of selective death may be much smaller under this model than under the previous model defined by (8.3) (Ohta 1978a). In general the reduction of genetic load by such modifications of the fitness function is more pronounced than the reduction of the effect of selection, and therefore the genetic load seems to be model dependent.

ii) Negative selection

It is possible that the gene diversity of multigene families is selected against. This is another approach to the mechanism of eliminating deleterious mutations from the multigene family (see section 8.3). It is also possible that uniformity of gene members may be preferable for a proper function of repeated genes with very homogeneous members such as histon gene families. A simple model for fitness may be the following.

$$W_i = 1 - s'(1 - c_i)$$

(8.12)

where s' is the selection coefficient, assumed to be positive. Under this model, the identity coefficient within a population is increased and genetic differentiation between isolated families may be decelerated, as shown below.

$$\Delta C_{w1} = \frac{s'\sigma_{w11}^2}{\bar{W}} \tag{8.13}$$

$$\Delta C_{w2} = \frac{2s'\sigma_{w12}}{\bar{W}} \tag{8.14}$$

and

$$\Delta C_b = \frac{2s'\sigma_{w1b}}{\bar{W}} \tag{8.15}$$

At the moment, the formulations are only for the change of identity coefficients in one generation and the general solution is not obtained. This is the area which needs future investigation. At any rate, natural selection should not be totally neglected in the discussion of gene diversity of multigene families. The analyses in chapter 6 on immunoglobulin diversity may need further consideration from this standpoint. If positive selection for immunoglobulin diversity operates on the hypervariable regions and negative selection, on the framework regions, the effect would be to increase the rate of evolutionary change at the hypervariable region and to decrease it at the framework regions. As to positive selection, the difficulty is that natural selection recognizes the whole variable region rather than the individual amino acid site as a unit to measure diversity. If it is possible to express variable region diversity as a sum of the amino acid site diversities over the sites in the proper region, the selection coefficient for the whole variable region would be the sum of the selection coefficients at individual amino acid sites, although linkage relationships further complicate the situation. Then selection coefficients at the amino acid sites should be extremely small and the effect may be indistinguishable from that of a high mutation rate. On the other hand, negative selection is more obviously at work in which definitely deleterious mutations that disturb basic immunoglobulin structure result in non-functional genes and are eventually eliminated before such mutations spread beyond a certain frequency in the family.

8.5 Possible other examples to apply the theory

Although there are not many examples of multigene families with such detailed information as immunoglobulin or ribosomal RNA gene families, several interesting cases are known. As a possibility, the

most attractive example is the major histocompatibility complex which
now represents a very active research subject in immunology (Klein
1979). The silkworm chorion genes that produce egg shell proteins
very likely constitute a multigene family (Kafatos et al. 1977,
Goldsmith and Basehoar 1978). In a broad definition of a multigene
family, the fibroin gene of silkworm also forms a multigene family;
the single large fibroin gene with internal repetition may be regarded
as a family made of many repeating units of two or more amino acid
sites (Lucas et al. 1957, Sprague et al. 1979). In this section, I
shall briefly discuss application of the theory to such genetic
systems.

   The major histocompatibility complex is a genetic system which
recognizes self and non-self substance and is known for its extensive
polymorphism. The system is as complex as its name implies, and
includes several genetic regions. The best studied systems are the
H-2 region in mouse and the HLA region in man. Each region of the
system is highly polymorphic and there are two main hypotheses to
explain this polymorphism; i) each region is a single locus at which
natural selection or high mutation rate or both are responsible for
high variability (Klein 1979), and ii) each region contains a cluster
of many loci (a multigene family), of which only one would be expressed (Bodmer 1973, Silver and Hood 1976). In order to make
theoretical predictions under the latter hypothesis, one needs the
present theory on genetic variation in multigene families. Fortunately, evolution and variation of some regions of the histocompatibility complex may now be measured by amino acid sequences; the
direct gene product (see Cunningham 1977, Klein 1979 for review).
Table 2 of Klein (1979) gives the alignment of the NH_2-terminal amino
acid sequences (28 sites) of H-2K and H-2D of mouse, HLA-A and HLA-B
of man and a few others. By using H-2K and HLA-B sequences because
of the large amount of data for them compared to other regions, I
estimate the identity coefficient for amino acids within a species
(C_{w2}) to be 0.89, and that between man and mouse (C_b), 0.64. These
values are very similar to the identity coefficients of immunoglobulin variable regions (see Chapter 6). Here the number of genes
per family (or per region in the present case) is expected to be
smaller in a histocompatibility region family than in a variable
region family of immunoglobulin. By comparing to the identity coefficient within a subfamily of immunoglobulins (Table 6.3, page 83),
the family size (number of genes in a region of major histocompatibility complex) would be about the same as that of a subfamily of

immunoglobulin, provided that the basic mutation rate is the same in both systems. Also, from equation (5.7), $-\ln(C_b/C_{w2}) = 0.33$ gives the estimated value of 2vt, in which, $t \approx 7 \times 10^7$ years, is the time since divergence of man and mouse, and v is the mutation rate per amino acid site per year which is equal to the rate of amino acid substitution in evolution in our model. Therefore $v \approx 2.36 \times 10^{-9}$ which is about twice as high as the evolutionary rate of the hemoglobin but is lower than that of the fibrinopeptides and is about the same as that of lactalbumin (Dayhoff 1972). At any rate, the extensive polymorphisms and the pattern of divergence between the species of the histocompatibility genes are in accord with the prediction of multigenic model.

On the other hand, if a region like H2-K or HLA-B consists of a single locus, one must assume strong natural selection. This is because, under the assumption of selective neutrality, the mutation rate must be at least one order of magnitude higher than ordinary protein loci, like that of hemoglobin, to have $C_{w2} = 0.89$, yet the mutation rate is only twice as high as that of hemoglobin as estimated above. Now, symmetric overdominance would be a most efficient type of diversifying selection. Here again high identity between the species, $C_b = 0.64$, as compared to the within-species identity ($C_{w2} = 0.89$) would contradict the prediction of natural selection if overdominance works at every amino acid site uniformly over the sequence, because differentiation should be more rapid under such a model. Thus, I suggest that diversifying selection is effective only at some limited sites and identity is kept high between the species at the remaining sites. A detailed study of this model will be published elsewhere. With available data, it seems to be impossible to discriminate between the multigene model and the single locus hypothesis with diversifying selection at restricted sites.

The chorion genes of silkworm provide another interesting system. It is almost certain that the genes form a multigene family (Goldsmith and Basehoar 1978). In this case, a large number of slightly different proteins are produced at certain stages of the development. Unfortunately, however, gene identity is difficult to estimate with available data at the moment. Fibroin gene of silkworm is interesting for its length heterogeneity (Sprague et al. 1979). Alleles of different strains are different in size. Apparently, unequal crossing-over is occurring fairly frequently in this gene region. Still another example is the androgen-regulated major urinary proteins of mouse (Hastie et al. 1979). They are synthesized

in large amounts in mouse liver and excreted into the urine. Hastie
et al. (1979) estimate that there are about 15 genes in a genome.
The corresponding gene of rat however is apparently coded by a
single locus. Thus, gene number should have increased in the recent
past. When more data are available on gene identity of these systems,
the theory developed in previous chapters would be a useful tool to
explain underlying mechanisms.

8.6 Multigene family of small size

Previous analyses are concerned with multigene families of large
sizes such as those of immunoglobulin variable regions and ribosomal
RNA genes. However, a gene family with only a few members per genome
seems to be fairly common. For example, Zimmer et al. (1980) report,
based on the pattern of restriction enzyme maps, that each genome of
human and some other primates generally contains two copies of the
hemoglobin α gene. However, these two gene copies must be very
similar to each other, as inferred from the level of amino acid identity
of their gene products in each species (Dayhoff 1972). This suggests
that the coincidental evolution, if such exists, should be rapid in
the hemoglobin α gene family. Here, an important point to note is
that the coincidental evolution is not merely a result of a single
duplication event, but is caused by continuous process of duplication
and deletion presumably through unequal crossing-over. It must be
essentially the same as in the case of multigene families consisting
of a large number of repeated genes. The analyses on mutant dynamics
for a population of gene families with a few copies in one genome
are highly desirable, which consider continuous occurrence of unequal
crossing-over as well as random genetic drift and mutation. Theoretical studies on a pair of duplicated genes in a population have
only started recently (Allendorf et al. 1975, Ferris and Whitt 1977,
Bailey et al. 1978, Kimura and King 1979, Takahata and Maruyama 1979).
All these studies treat the case where gene duplication is caused by
polyploidization followed by rapid differentiation of the duplicate
copies. Therefore, unequal crossing-over was not needed to be taken
into account. However, a treatment similar to that for multigene
families of large sizes is required when tandemly duplicated genes
are involved.

In addition to the above considerations, it is reasonable to
assume that the rate of unequal crossing-over decreases as similarity
between members diminishes (Seidman et al. 1978, Zimmer et al. 1980).
The rate may also depend upon the presence of internal repetitions

which may be created in the spacer region (Smith 1976, Fedoroff 1979). Thus, in order to fully understand the evolution of repeated gene families, highly complicated analyses will be required.

8.7 Concluding remarks

Owing to the remarkable progress of eukaryote molecular biology, the organization of genetic material has been much clarified in recent years. Some of such findings are revolutionary to basic biological concepts. With the technique of cloning and sequencing of DNA, it is now known that many genes of eukaryotes are quite large, and are usually split into pieces and contain several "introns" which are cut off and discarded during the processing of messenger RNA. At the very top of such investigations is the unraveling of the organization of immunoglobulin gene families which has been discussed in the previous chapters.

Evolutionary theory should be no exception from such a scientific revolution. In 1968, Kimura (1968) proposed a neutral theory of molecular evolution which states that the majority of amino acid substitutions in evolution must be neutral with respect to natural selection and due to random genetic drift at reproduction. In the next year, King and Jukes (1969) advocated the theory from the more biochemical standpoint in the name of "non-Darwinian evolution". Since this hypothesis is totally against the neo-Darwinian view of evolution, it met strong criticisms and objections in the sub-sequent ten years (see Kimura 1979 for review). Although the original theory needed a few modifications (Ohta 1974), it has survived and much data have suggested its correctness. To me, the neutral theory is a beginning of reformulations of some of the basic models of Darwinian natural selection.

In his review of the influential book by Lewontin (1974) "The Genetic Basis of Evolutionary Change", Robertson (1975) remarked that the evolution of repeated gene families is totally outside the scope of the book. The statement has stimulated me to study this subject more intensely. Not only for repeated gene families, but also for ordinary gene loci, duplication or deletion of genetic material seems to be more common and important for evolution than previously thought. It appears that, by cloning and sequencing of DNA, new cases are being revealed in which genes are duplicated in the genome, although they were thought to be singly represented (Nakanishi et al. 1979, Hastie et al. 1979, Royal et al. 1979). The correct understanding of the organization of immunoglobulin gene families has also been possible only after cloning and sequencing of the DNA region of this gene family (Marx 1978).

Excess amounts of DNA in a genome of higher organisms as compared with the total number of gene loci (Muller 1950) is becoming understandable with new information of large gene size and redundancy of some genes. It has been argued that the part of genome DNA which codes proteins and carries essential genetic information should be only several percent of the total DNA (e. g. King and Jukes 1969, Ohta and Kimura 1971). This thesis is also more realistic with new information than before. At any rate, duplication, deletion and recombination of genetic information seem to be playing basic roles for progressive evolution of higher organisms, which perhaps can only occur in a flexible genome with non-informational regions and partial redundancy. Dayhoff (1978, page 10) conjectures that all proteins in humans, which she estimated to be about 50,000, will be grouped into 500 superfamilies, each containing an average of 100 proteins. This also implies the importance of gene duplication in evolution.

A final remark concerns the evolution of recombination recently discussed by Williams (1975), Maynard Smith (1978) and others. As compared with the classical model of Mendelian genes, the rate of recombination has a direct effect on evolution and variation of multigene families. That is by means of recombining and exchanging mutant genes between the different gene families. Equation (3.22) tells how recombination rate (β') directly influences the gene diversity of the family. At least one important genetic system, i. e. immunoglobulin gene families, owes its enormous diversity to recombination, in addition to combinatorial use of genetic information through ontogeny. Thus the problem of the evolution of recombination now looks in a new direction as does the assessment of unequal crossing-over.

REFERENCES

Allendorf, F. W., F. M. Utter and B. P. May. (1975). Gene duplication within the family Salmonidae: II. Detection and deter-
mination of the genetic control of duplicate loci through
inheritance studies and the examination of populations. in
Isozymes, ed. C. L. Markert (Academic, New York), Vol. 4,
pp. 415-432.

Bailey, G. S., R. T. M. Poulter and P. A. Stockwell. (1978). Gene
duplication in tetraploid fish: Model for gene silencing at
unlinked duplicate loci. Proc. Nat. Acad. Sci. USA 75:
5575-5579.

Birky, C. W. Jr. (1978). Transmission genetics of mitochondria and
chloroplasts. Ann. Rev. Genetics 12: 471-512.

Black, J. A. and D. Gibson. (1974). Neutral evolution and immunoglobulin diversity. Nature 250: 327-328.

Bodmer, W. F. (1973). A new genetic model for allelism at histocompatibility and other complex loci: Polymorphism for control
of gene expression. Transplant. Proc. 5: 1471-1476.

Britten, R. J. and D. E. Kohne. (1968). Repeated sequences in DNA.
Science 161: 529-540.

Brown, D. D. and K. Sugimoto. (1973). 5SDNAs of Xenopus laevis and
Xenopus mulleri: Evolution of a gene family. J. Mol. Biol.
78: 397-415.

Brown, D. D., P. C. Wensink and E. Jordan. (1971). A comparison of
the ribosomal DNAs of Xenopus laevis and Xenopus mulleri: the
evolution of tandem genes. J. Mol. Biol. 63: 57-73.

Callan, H. G. (1967). The organization of genetic units in chromosomes. J. Cell Sci. 2: 1-7.

Capra, J. D. and A. B. Edmundson. (1977). The antibody combining
site. Scientific American 236(1): 50-59.

Chakraborty, R., P. A. Fuerst and M. Nei. (1978). Statistical
studies on protein polymorphism in natural populations II.
Gene differentiation between populations. Genetics 88: 367-390.

Clayton, G. A. and A. Robertson. (1957). An experimental check on
quantitative genetical research. Jour. Genet. 55: 152-170.

Cold Spring Harbor Laboratory (1977). Proc. Cold Spring Harbor
Symposia on Quantitative Biology, Vol. 41, "Origins of Lymphocyte
Diversity".

Crow, J. F. and M. Kimura. (1970). An Introduction to Population Genetics Theory. Harper & Row, New York.

Cunningham, B. A. (1977). The structure and function of histocompatibility antigens. Scientific American 237(4): 96-107.

Davidson, E. H. and R. J. Britten. (1971). Repetitive and non-repetitive DNA sequences and a speculation on the origin of evolutionary novelty. Q. Rev. Biol. 46: 111-138.

Dayhoff, M. O. (1972). Atlas of Protein Sequence and Structure 1972. National Biomedical Research Foundation, Silver Spring, Maryland.

Dayhoff, M. O. (1978). Atlas of Protein Sequence and Structure, Vol. 5, supplement 3. National Biomedical Research Foundation, Georgetown Univ. Med. Center, Washington.

Dover, G. (1978). DNA conservation and speciation: adaptive or accidental? Nature 272: 123-124.

Fedoroff, N. V. (1979). On spacers. Cell 16: 697-710.

Fedoroff, N. V. and D. D. Brown. (1978). The nucleotide sequence of the repeating unit in the oocyte 5S ribosomal DNA of Xenopus laevis. Proc. Cold Spring H. Symp. 42: 1195-1200.

Feller, W. (1957). An Introduction to Probability Theory and its Applications. Vol. 1. John Wiley, New York.

Ferris, S. D. and G. S. Whitt. (1977). Loss of duplicate gene expression after polyploidisation. Nature 265: 258-260.

Finnegan, D. J., G. M. Rubin, M. W. Young and D. S. Hogness. (1978). Repeated gene families in Drosophila melanogaster. Proc. Cold Spring H. Symp. Quant. Biol. 42: 1053-1063.

Fisher, R. A. (1930). The Genetical Theory of Natural Selection. Clarendon Press, Oxford.

Fitch, W. M. (1973). Aspects of molecular evolution. Ann. Review of Genetics 7: 343-380.

Frankham, R., Briscoe, D. A. and Nurthen, R. K. (1978). Unequal crossing-over at the rRNA locus as a source of quantitative genetic variation. Nature 272, 80-81.

Fry, K. and W. Salser. (1977). Nucleotide sequence of HS-α satellite DNA from kangaroo rat Dipodomys ordii and characterization of similar sequences in other rodents. Cell 12: 1069-1084.

Gally, J. A. and G. M. Edelman. (1972). The genetic control of immunoglobulin synthesis. Ann. Rev. Genetics 6: 1-46.

Goldsmith, M. R. and G. Basehoar. (1978). Organization of the chorion genes of Bombyx mori, a multigene family. I. Evidence for linkage to the second chromosome. Genetics 90: 291-310.

Hastie, N. D., W. A. Held and J. J. Toole. (1979). Multiple genes coding for the androgen-regulated major urinary proteins of mouse. Cell 17: 449-457.

Hennig, W. and P. M. B. Walker. (1970). Variations in the DNA from two rodent families (Cricetidae and Muridae) Nature 225: 915-919.

Hill, W. G. (1975). Linkage disequilibrium among multiple neutral alleles produced by mutation in finite population. Theor. Pop. Biol. 8: 117-126.

Hill, W. G. and A. Robertson. (1968). Linkage disequilibrium in finite populations. Theor. Appl. Genetics 38: 226-231.

Hood, L., J. H. Campbell and S. C. R. Elgin. (1975). The organization, expression, and evolution of antibody genes and other multigene families. Ann. Rev. Genetics 9: 305-353.

Jerne, N. K. (1977). The common sense of immunology. Proc. Cold Spring H. Symp. on Quant. Biol. 41: 1-4.

Kabat, E. A., T. T. Wu and H. Bilofsky. (1976). Variable regions of immunoglobulin chains. Medical Computer Systems, Bolt, Beranek and Newman, Cambridge, Mass.

Kabat, E. A., T. T. Wu and H. Bilofsky. (1979). Evidence supporting somatic assembly of the DNA segments (minigenes), coding for the framework, and complementarity-determining segments of immunoglobulin variable regions. J. Exp. Med. 149: 1299-1313.

Kafatos, F. C., J. C. Regier, G. D. Mazur, M. R. Nadel, H. M. Blau, W. H. Petri, A. R. Wyman, R. E. Gelinas, P. B. Moore, M. Paul, A. Efstratiadis, J. N. Vournakis, M. R. Goldsmith, J. R. Hunsley, B. Baker, J. Nardi and M. Koehler. (1977). The eggshell of insects: differentiation-specific proteins and the control of their synthesis and accumulation during development. "Results and Problems in Cell Differentiation" Vol. 8, pp. 45-145. W. Beerman, ed. Springer-Verlag, N. Y. & Berlin.

Kimura, M. (1964). Diffusion models in population genetics. Jour. Applied Probability 1: 177-232.

Kimura, M. (1968). Evolutionary rate at the molecular level. Nature 217: 624-626.

Kimura, M. (1979). The neutral theory of molecular evolution. Scientific American 241(5): 94-104.

Kimura, M. and J. F. Crow. (1964). The number of alleles that can be maintained in a finite population. Genetics 49: 725-738.

Kimura, M. and J. L. King. (1979). Fixation of a deleterious allele
    at one of two "duplicate" loci by mutation pressure and random
    drift. Proc. Nat. Acad. Sci. USA 76: 2858-2861.
Kimura, M. and T. Ohta. (1969a). The average number of generations
    until fixation of a mutant gene in a finite population.
    Genetics 61: 763-771.
Kimura, M. and T. Ohta. (1969b). The average number of generations
    until extinction of an individual mutant gene in a finite
    population. Genetics 63: 701-709.
Kimura, M. and T. Ohta. (1971a). Protein polymorphism as a phase
    of molecular evolution. Nature 229: 467-469.
Kimura, M. and T. Ohta. (1971b). Theoretical Aspects of Population
    Genetics. Princeton University Press, Princeton.
Kimura, M. and T. Ohta. (1974). On some principles governing
    molecular evolution. Proc. Nat. Acad. Sci. USA 71: 2848-2852.
Kimura, M. and T. Ohta. (1979). Population genetics of multigene
    family with special reference to decrease of genetic correlation
    with distance between gene members on a chromosome. Proc. Nat.
    Acad. Sci. USA 76: 4001-4005.
King, J. L. and T. H. Jukes. (1969). Non-Darwinian evolution:
    Random fixation of selectively neutral mutations. Science
    164: 788-798.
Klein, J. (1979). The major histocompatibility complex of the
    mouse. Science 203: 516-521.
Krystal, Mark and N. Arnheim. (1978). Length heterogeneity in a
    region of human ribosomal gene spacer is not accompanied by
    extensive population polymorphism. J. Mol. Biol. 126: 91-104.
Leder, P., T. Honjo, J. Seidman and D. Swan. (1977). Origin of
    Immunoglobulin Gene Diversity: The Evidence and a Restriction-
    modification Model. Proc. Cold Spring Harbor Symp. on Quant
    Biol. 41: 855-862.
Lewontin, R. C. (1974). The genetic basis of evolutionary change.
    Columbia University Press, New York and London.
Li, W-H. and M. Nei. (1975). Drift variances of heterozygosity and
    genetic distance in transient states. Genet. Res. 25: 229-248.
Lucas, F., J. T. B. Shaw and S. G. Smith. (1957). The amino acid
    sequence in a fraction of the fibroin of Bombyx mori.
    Biochem. J. 66: 468-479.
Marx, J. L. (1978). Antibodies (I): New information about gene
    structure. Science 202: 298-299. Antibodies (II): Another look
    at the diversity problem. Science 202: 412-415.

Mather, K. (1949). Biometric Genetics: The Study of Continuous Variations. Dover Publications Inc., New York.

Maynard Smith, J. (1978). The Evolution of Sex. Cambridge Univ. Press, Cambridge and New York.

Muller, H. J. (1950). Our load of mutations. Amer. J. Hum. Genet. 2: 111-176.

Musich, P. R., F. L. Brown and J. J. Maio. (1978). Mammalian repetitive DNA and the subunit structure of chromatin. Proc. Cold Spring H. Symp. 42: 1147-1160.

Nakanishi, S., A. Inoue, T. Kita, N. Nakamura, A. C. Y. Chang, S. N. Cohen and S. Numa. (1979). Nucleotide sequence of cloned cDNA for bovine corticotropin-β-lipotropin precursor. Nature 278: 423-427.

Nei, M. (1972). Genetic distance between populations. Amer. Nat. 106: 283-292.

Nei, M. (1975). Molecular Population Genetics and Evolution. North-Holland and Elsevier, Amsterdom·Oxford.

Nei, M. and A. K. Roychoudhury. (1974). Sampling variances of heterozygosity and genetic distance. Genetics 76: 379-390.

Novotný, J., A. Vitek and F. Franěk. (1977). Analysis of inter-species variability of lambda chains reveals differences in domain tertiary structure. J. Mol. Biol. 113: 711-718.

Ohno, S. (1970). Evolution by Gene Duplication. Springer, Berlin.

Ohta, T. (1974). Mutational pressure as the main cause of molecular evolution and polymorphisms. Nature 252: 351-354.

Ohta, T. (1976). A simple model for treating the evolution of multigene families. Nature 263: 74-76.

Ohta, T. (1977). On the gene conversion model as a mechanism for maintenance of homogeneity in systems with multiple genomes. Genet. Res. 30: 89-91.

Ohta, T. (1978a). Theoretical study on genetic variation in multi-gene families. Genet. Res. 31: 13-28.

Ohta, T. (1978b). Theoretical population genetics of repeated genes forming a multigene family. Genetics 88: 845-861.

Ohta, T. (1978c). Sequence variability of immunoglobulins considered from the standpoint of population genetics. Proc. Nat. Acad. Sci. USA 75: 5108-5112.

Ohta, T. (1979). An extension of a model for the evolution of multigene families by unequal crossing-over. Genetics 91: 591-607.

Ohta, T. (1980a). Amino acid diversity of immunoglobulins as a product of molecular evolution. J. Mol. Evol. (in press).

Ohta, T. (1980b). Linkage disequilibrium between amino acid sites in immunoglobulin genes and other multigene families. Genet. Res. (submitted).

Ohta, T. and M. Kimura. (1969). Linkage disequilibrium due to random genetic drift. Genet. Res. 13: 47-55.

Ohta, T. and M. Kimura. (1971a). Linkage disequilibrium between two segregating nucleotide sites under the steady flux of mutations in a finite population. Genetics 68: 571-580.

Ohta, T. and M. Kimura. (1971b). Functional organization of genetic material as a product of molecular evolution. Nature 233: 118-119.

Perelson, A. S. and G. I. Bell. (1977). Mathematical models for the evolution of multigene families by unequal crossing over. Nature 265: 304-310.

Ritossa, F., C. Malva, E. Boncinelli, F. Graziani and L. Polito. (1971). The first steps of magnification of DNA complementary to ribosomal RNA in Drosophila melanogaster. Proc. Nat. Acad. Sci. USA 68: 1580-1584.

Robertson, A. (1975). Review on "The Genetic Basis of Evolutionary Change" by R. C. Lewontin. Nature 254: 367.

Royal, A., A. Garapin, B. Cami, F. Perrin, J. L. Mandel, M. LeMeur, F. Brégégègre, F. Cannon, J. P. LePennec, P. Chambon and P. Kourilsky. (1979). The ovalbumin gene region: common features in the organization of three genes expressed in chicken oviduct under hormonal control. Nature 279: 125-132.

Sakano, H., K. Hüppi, G. Heinrich and S. Tonegawa. (1979). Sequences at the somatic recombination sites of immunoglobulin light-chain genes. Nature 280: 288-294.

Salser, W., S. Bowen, D. Browne, F. El Adli, N. Fedoroff, K. Fry, H. Heindell, G. Paddock, R. Poon, B. Wallace and P. Whitcome. (1976). Investigation of the organization of mammalian chromosomes at the DNA sequence level. Fed. Proc. 35: 23-35.

Seidman, J. G., A. Leder, M. N. B. Norman and P. Leder. (1978). Antibody diversity. Science 202: 11-17.

Seidman, J. G., E. E. Max and P. Leder. (1979). A κ-immunoglobulin gene is formed by site-specific recombination without further somatic mutation. Nature 280: 370-375.

Silver, J. and L. Hood. (1976). Structure and evolution of tran-
    plantation antigens: Partial amino acid sequences of H-2K and
    H-2D alloantigens. Proc. Nat. Acad. Sci. USA 73: 599-603.

Smith, G. P. (1974). Unequal crossover and the evolution of multi-
    gene families. Proc. Cold Spring Harbor Symp. Quant. Biol.
    38: 507-513.

Smith, G. P. (1976). Evolution of repeated DNA sequences by un-
    equal crossover. Science 191: 528-535.

Smith, G. P. (1977). The significance of hybridization kinetic
    experiments for theories of antibody diversity. Proc. Cold
    Spring Harbor Symp. Quant. Biol. 41: 863-875.

Southern, E. M. (1970). Base sequence and evolution of guinea-pig
    α-satellite DNA. Nature 227: 794-798.

Sprague, K. U., M. B. Roth, R. F. Manning and L. P. Gage (1979).
    Alleles of the fibroin gene coding for proteins of different
    length. Cell 17: 407-413.

Stewart, F. M. (1976). Variability in the amount of heterozygosity
    maintained by neutral mutations. Theor. Pop. Biol. 9: 188-201.

Takahata, N. and T. Maruyama. (1979). Polymorphism and loss of
    duplicate gene expression: A theoretical study with application
    to tetraploid fish. Proc. Nat. Acad. Sci. USA 76: 4521-4525.

Tartof, K. D. (1975). Redundant genes. Ann. Rev. Genetics
    9: 355-385.

Tartof, K. and I. B. Dawid. (1976). Similarities and differences
    in the structure of X and Y chromosome rRNA genes of Drosophila.
    Nature 263: 27-30.

Tonegawa, S., N. Hozumi, G. Matthyssens and R. Schuller. (1977).
    Somatic changes in the content and context of immunoglobulin
    genes. Proc. Cold Spring H. Symp. Quant. Biol. 41: 877-889.

Tonegawa, S., A. M. Maxam, R. Tizard, O. Bernard and W. Gilbert.
    (1978). Sequence of a mouse germline gene for a variable
    region of an immunoglobulin light chain. Proc. Nat. Acad. Sci.
    USA 75: 1485-1489.

Valbuena, O., K. B. Marcu, M. Weigert and R. P. Perry. (1978).
    Multiplicity of germline genes specifying a group of related
    mouse κ chains with implications for the generation of immuno-
    globulin diversity. Nature 276: 780-784.

Walker, P. M. B. (1971). Repetitive DNA in higher organisms. Prog.
    Biophys. Mol. Biol. 23: 145-190.

Watson, J. D. (1976). Molecular Biology of the Gene, 3rd edition.
    W. A. Benjamin, Inc., Menlo Park, Calif., U. S. A.

Weigert, M., L. Gatmaitan, E. Loh, J. Schilling and L. Hood. (1978).
    Rearrangement of genetic information may produce immunoglobulin
    diversity. Nature 276: 785-790.

Weigert, M. and R. Riblet. (1977). Genetic control of antibody
    variable regions. Proc. Cold Spring H. Symp. Quant. Bio..
    41: 837-846.

Wellauer, P. K., R. H. Reeder, D. Carroll, D. D. Brown, A. Deutch,
    T. Higashinakagawa and I. B. Dawid. (1974). Amplified ribosomal
    DNA from Xenopus leavis has heterogeneous spacer lengths.
    Proc. Nat. Acad. Sci. USA 71: 2823-2827.

Wellauer, P. K., R. H. Reeder, I. B. Dawid and D. D. Brown. (1976).
    The arrangement of length heterogeneity in repeating units of
    amplified and chromosomal ribosomal DNA from Xenopus laevis.
    J. Mol. Biol. 105: 487-505.

Whittaker, E. T. and G. N. Watson. (1935). A Course of Modern
    Analysis, 4th Ed. Cambridge Univ. Press, Cambridge.

Williams, G. C. (1975). Sex and Evolution. Princeton Univ. Press,
    Princeton.

Wright, Sewall. (1969). Evolution and the Genetics of Populations
    Vol. 2. The Theory of Gene Frequencies. The Univ. of Chicago
    Press, Chicago and London.

Wright, Sewall. (1977). Evolution and the Genetics of Populations.
    Vol. 3. Experimental Results and Evolutionary Deductions.
    The Univ. of Chicago Press, Chicago and London.

Wu, T. T. and E. A. Kabat. (1970). An analysis of the sequences of
    the variable regions of Bence Jones proteins and myeloma light
    chains and their implications for antibody complementarity.
    J. Experimental Medicine 132: 211-250.

Yamamoto, M. and G. L. G. Miklos. (1978). Genetic studies on
    heterochromatin in Drosophila melanogaster and their implications
    for function of satellite DNA. Chromosoma 66: 71-98.

Zimmer, E. A., S. L. Martin, S. M. Beverley, Y. W. Kan and A. C.
    Wilson. (1980). Coincidental evolution of globin genes in
    relation to intron length and intergenic distance.
    Submitted for publication.

SUBJECT INDEX

Bio-mathematics

Managing Editors: K. Krickeberg, S. A. Levin

Springer-Verlag
Berlin
Heidelberg
New York

Volume 8

A. T. Winfree

The Geometry of Biological Time

1979. Approx. 290 figures. Approx. 580 pages
ISBN 3-540-09373-7

The widespread appearance of periodic patterns in nature reveals that many living organisms are communities of biological clocks. This landmark text investigates, and explains in mathematical terms, periodic processes in living systems and in their non-living analogues. Its lively presentation (including many drawings), timely perspective and unique bibliography will make it rewarding reading for students and researchers in many disciplines.

Volume 9

W. J. Ewens

Mathematical Population Genetics

1979. 4 figures, 17 tables. XII, 325 pages
ISBN 3-540-09577-2

This graduate level monograph considers the mathematical theory of population genetics, emphasizing aspects relevant to evolutionary studies. It contains a definitive and comprehensive discussion of relevant areas with references to the essential literature. The sound presentation and excellent exposition make this book a standard for population geneticists interested in the mathematical foundations of their subject as well as for mathematicians involved with genetic evolutionary processes.

Volume 10

A. Okubo

Diffusion and Ecological Problems: Mathematical Models

1980. 114 figures. XIII, 254 pages
ISBN 3-540-09620-5

This is the first comprehensive book on mathematical models of diffusion in an ecological context. Directed towards applied mathematicians, physicists and biologists, it gives a sound, biologically oriented treatment of the mathematics and physics of diffusion.

Journal of

Mathematical Biology

ISSN 0303-6812   Title No. 285

Editorial Board:
H.T.Banks, Providence, RI; **H.J.Bremermann,**
Berkeley, CA; **J.D.Cowan,** Chicago, IL; **J.Gani,**
Canberra City; **K.P.Hadeler** (Managing Editor),
Tübingen; **S.A.Levin** (Managing Editor), Ithaca, NY;
D.Ludwig, Vancouver; **L.A.Segel,** Rehovot; **D.Varjú,**
Tübingen

Advisory Board: M. A. Arbib, W.Bühler, B.D.Coleman,
K. Dietz, F. A. Dodge, P. C. Fife, W. Fleming, D. Glaser,
N. S. Goel, S. P. Hastings, W. Jäger, K. Jänich, S. Karlin,
S. Kauffman, D. G. Kendall, N. Keyfitz, B. Khodorov,
J. F. C. Kingman, E. R. Lewis, H. Mel, H. Mohr,
E. W. Montroll, J. D. Murray, T. Nagylaki, G. M. Odell,
G. Oster, L. A. Peletier, A. S. Perelson, T. Poggio,
K. H. Pribram, J. M. Rinzel, S. I. Rubinow, W. v. Seelen,
W. Seyffert, R. B. Stein, R. Thom, J. J. Tyson

Springer-Verlag
Berlin
Heidelberg
New York

The **Journal of Mathematical Biology** publishes papers
in which mathematics leads to a better understanding
of biological phenomena, mathematical papers inspired
by biological research and papers which yield new experimental data bearing on mathematical models. The
scope is broad, both mathematically and biologically
and extends to relevant interfaces with medicine,
chemistry, physics and sociology. The editors aim to
reach an audience of both mathematicians and
biologists.

Subscription information and sample copy
upon request.

Lecture Notes in Biomathematics

This series reports new developments in biomathematics research and teaching – quickly, informally and at a high level. The type of material considered for publication includes:

1. Preliminary drafts of original papers and monographs

2. Lectures on a new field or presentations of new angles in a classical field

3. Seminar work-outs

4. Reports of meetings, provided they are

   a) of exceptional interest and

   b) devoted to a single topic.

Texts which are out of print but still in demand may also be considered if they fall within these categories.

The timeliness of a manuscript is more important than its form, which may be unfinished or tentative. Thus, in some instances, proofs may be merely outlined and results presented which have been or will later be published' elsewhere. If possible, a subject index should be included. Publication of Lecture Notes is intended as a service to the international scientific community, in that a commercial publisher, Springer-Verlag, can offer a wide distribution of documents which would otherwise have a restricted readership. Once published and copyrighted, they can be documented in the scientific literature.

Manuscripts

Manuscripts should be no less than 100 and preferably no more than 500 pages in length.
They are reproduced by a photographic process and therefore must be typed with extreme care. Symbols not on the typewriter should be inserted by hand in indelible black ink. Corrections to the typescript should be made by pasting in the new text or painting out errors with white correction fluid. Authors receive 75 free copies and are free to use the material in other publications. The typescript is reduced slightly in size during reproduction; best results will not be obtained unless the text on any one page is kept within the overall limit of 18 x 26.5 cm (7 x 10½ inches). On request, the publisher will supply special paper with the typing area outlined.

Manuscripts in English, German or French should be sent to Dr. Simon Levin, 235 Langmuir, Cornell University, Ithaca, NY 14850/USA or directly to Springer-Verlag Heidelberg.

Springer-Verlag, Heidelberger Platz 3, D-1000 Berlin 33
Springer-Verlag, Neuenheimer Landstraße 28–30, D-6900 Heidelberg 1
Springer-Verlag, 175 Fifth Avenue, New York, NY 10010/USA

ISBN 3-540-09998-0
ISBN 0-387-09998-0